Internal Components

MITCHELL'S BUILDING SERIES

Internal Components

ALAN BLANC

Longman
Scientific &
Technical

Longman Scientific & Technical
Longman Group UK Limited
Longman House, Burnt Mill, Harlow
Essex, CM20 2JE, England
and Associated Companies throughout the world

First published 1994

ISBN 0-582-21257X

British Library Cataloguing in Publication Data
A catalogue entry for this title is available from the British Library.

Set by 4 in Compugraphic Times and Melior
Produced by Longman Singapore Publishers (Pte) Ltd
Printed in Singapore

Contents

Acknowledgements

The work on *Internal Components* would not have been possible without the enthusiastic support of Basil Wilby of Longman, the in-house team at Harlow, and the new illustrator. Acknowledgements need to be paid to Thelma Nye of Batsford for her very able editorial direction over many years and to Jean Marshall who prepared much of the illustrative material that has been retained.

Many building suppliers have been approached for up-to-date details, and acknowledgement for their assistance is given alongside the figures and photographs. Thanks are also due to Barric Evans, technical editor of *The Architects' Journal*, for his vigorous criticism and valuable advice on the early draft.

Copies of British Standards Institution publications may be obtained from BSI, Linford Wood, Milton Keynes, MK14 6LE.

Finally, warmest thanks are due to Sylvia Blanc and Ruth Lush who converted sheaves of notes and revisions into typescript, with special thanks to Sylvia who stayed the course and finally prepared the index, a true labour of love.

Alan Blanc

Preface

The tragic death of Derek Osbourn severed his invaluable help with the Mitchell's Building Series that spanned the past decades. The *Components* volume as updated by Osbourn in 1989 had retained the style and presentation from the original 1971 edition by Harold King. This in turn had preserved many of the illustrations that dated back to the earlier work of Denzil Neal.

At the time of Derek Osbourn's death, ideas for the restructuring the Mitchell's titles were already in hand. A significant advisory role has been played by Yvonne Dean, a teaching colleague of Osbourn's from the former Polytechnic of North London (PNL) (now the University of North London). The 1990s' editions extend the theoretical base by accompanying the 'reasons why' attitude with the traditional 'hands on' approach to construction technology.

There is still a considerable acknowledgement to the influence of Derek Osbourn and to his generosity in giving other architects or specialists a chance to participate in the technical writing or to share in the lecture work. Components formed a significant part of the lecture material prepared in the early 1980s which was tested with the live critical audiences of full- and part-time students at PNL. It should be explained that I, in those days at PNL, used to travel from the Holloway Road campus over to the North East London Polytechnic (now the University of East London) to give a series of technology lectures based upon Osbourn's syllabus. The pattern of instruction was conventional, with summaries and hand-out sketches, always with the intention that students would refer subsequently to textbooks for fuller detail. Similar methods were applied at the Polytechnic of Central London (now the University of Westminster) where Michael McEvoy, who has written this book's companion volume *External Components*, is now Co-ordinator of Technical Studies.

The starting point for much of the material has been the pages of *The Architects' Journal* (*AJ*), together with the supplements in the *AJ Focus*. Some of the rewriting has come full circle: my contribution to the Art of Construction series for the AJ in the early 1980s was included at the request of Derek Osbourn when he took over the principal's role at PNL.

A theme of my experience is the mixture of practice and teaching which brings a sharper focus to the needs of textbooks for the 1990s. To keep matters in reasonable proportion, two volumes have been prepared where only one existed previously, with a logical split to external and internal components. The regrettable reduction in lecture block hours within higher education makes book provision a critical issue and it is therefore welcome news that Longman is fully committed to developing the Mitchell's Building Series to the conditions of today.

1994   Alan Blanc

1 Component design

1.1 Introduction

The definition of a *Building component* is given in BS 6100 *Building and civil engineering terms* Part 1 *General and miscellaneous* subsection 1.5.1 : 1984 Co-ordination of dimensions, tolerances and accuracy. It is defined as a 'product manufactured as a distinct unit to serve a specific function or functions'. The relationship of these various components one to another, as well as the components themselves, contribute to the visual quality of the building and provide the basis for arriving at the art of construction. At present the art of construction requires a vast range of components since technology has not involved a monolithic form of building using one material (like ice-blocks for an igloo) nor a building process reliant upon a single technique, such as tree houses in Amazonia.

Industrialization has diversified the materials and assembly techniques employed. For example, a conventional house requires bricks, blocks or timber-framed panels for walls while precast concrete, steel or timber beams are used for floor and roof structures. Roof coverings will be slate, tile or sheet materials. Doors, windows and cladding trim, components in themselves, will be formed from profile components of timber, metal and/or plastic, not to mention glass and glazing units. Internal finishings will be plasterboard, sheeting from chipboard or hardboard with decorative surfaces made from laminates or with flooring using carpet, vinyl or clay tile. Glazed ware, metal or plastic form sanitary fittings. Flat-pack or fabricated joinery form kitchen cabinets and wardrobes. Kits also comprise the essential elements for staircase assembly. Ironmongery is another component packaged as sets of fittings and completing the *bricolage* that is assembled from builders' merchants for the domestic building market. It is a feature of residential construction that is commonplace in Western Europe, the USA, Canada, Australasia and wherever the marketing of industrial products are developed. The assembly or erection time spent on site is about a third of the total labour impact; the balance is spent on off-site manufacture and preparation. The economics of inflation has meant that standardized items are reduced in choice unless volume orders can be guaranteed. This is a very different scenario from that presented by Harold King in the 1971 edition of *Components and Finishes*[1] when industrialized building systems were seen as a total panacea. Inflation and lack of orders has seen off the viability of the systems approach. The domestic market is, however, one of the few openings in the UK where 'off the shelf' sales of building components still predominate.

The example of standardized domestic components has a long history in the building of London. Nicholas Barbon (1640−98), a speculator involved after the Great Fire of London, employed joinery features and staircases made to standard sizes and designs. This is a different notion from the pattern books put out by James Gibbs or Isaac Ware which were adapted by craftsmen to suit the site. Thomas Cubitt (1788−1855) took the notion of off-site preparation to the ultimate degree in traditional terms, his Pimlico workshops servicing the adjacent estate and the whole of the development in Belgravia.[2] The standard items became the catalogue elements illustrated in Victorian brochures and extended to doors, windows and trim (stone, timber and fibrous plaster) as well as balconies, balustrading, stairs and fireplace fronts, including mantles and surrounds and even picture surrounds and wall mirrors!

Designers have always been intrigued by the achievement of Sir Joseph Paxton in his production of factory-made components to achieve the seemingly impossible building programme at the Crystal Palace to house the 1851 Great Exhibition. Paxton had simply extended and rationalized a system of conservatory construction to cover 18 acres in the phenomenal time of 4 months. The fact that the wooden glazing bar components were inadequate and leaked like a sieve until replaced by metal bars in 1899 has largely been forgotten.[3] The seminal lesson from the Crystal Palace is the way mechanization of building manufacture away from the site can speed the process of construction. The other salient point is the degree of repetition achieved to ensure economy of production. The operational differences from Paxton's time to today are not only the greater scale and sophistication of the building industry in the late twentieth century but the way Codes of Practice today embrace every aspect of building performance. The performance aspect has made crucial changes to the way component specifications are prepared, the craft-based approach of times past giving way to a scientific basis for describing building achievement.

Another illustration of change is the way components are fabricated together in larger subcontracts, for example, rain screen façades that include wall framing, services, external cladding and glazed windows. There is also the direction anticipated by 'Archigram'[4] whereby buildings are put together from structural pods. This is a policy adopted for the sanitary accommodation in the service towers of the Lloyd's Building, City of London (1984–86), architects Richard Rogers and Partners. At a more humdrum level it is customary to employ factory-made service cores for hotel bathrooms and for upgrading of older housing. Industrialization is not exclusively concerned with factory production nor with changes of building method, but involves a new attitude to design and construction and to the way buildings are commissioned. This concerns client and user, designer or architect, administrator or project manager, manufacturer or subcontractor and general contractor. Industrialized processes break down the traditional divisions of responsibility and favour a mutually dependent organization. This is different from the traditional roles where design and construction are carried forward under separate management control. A typical example of the latter would be extensive repairs to a historic building and where conventional craft-based components such as brick and stone masonry are engaged.

The change in the way buildings are procured by clients today affects not only specification writing but the way technical information is prepared for contracts. 'Design and build' procedures where the client contracts these services from contractors break down the old divisions within the building industry, with the client and contractor in contract, but with design services directed through the professional advice of the architect, engineer or surveyor, or any of these professionals being the supervising officer for the work involved. Increasingly the construction team is led by the management contractor and together they devise subcontracts whereby the various elements of the building work are split into 'packages', namely labour and materials packages of work assigned by the main contractor. Typical examples are the building frame, roofing, wall cladding and the various services, also groups of components such as doors, joinery, partitions and staircases.

The designer's role will still be concerned with furnishing sufficient information for the specialists to price the work, but with the emphasis on the salient geometry and the performance requirements in use. The specification follows the same direction, leaving the specialist to arrive at the most economic method of manufacture to fulfil the performance criteria set down. The designer's role is in some ways more extensive than in times past since knowledge is expected of the 'state of the art' of construction in order that the latest technology can be applied. The specialist's role will lead their participation with workshop drawings of the eventual production, but these in turn have to be checked by the designers. The final installation rests between the main contractor and the specialist.

This process is very different from the craft-based building contracts of times past when designers sought tenders for subcontracts and then nominated the successful firm for the work or drew up an agreement by direct negotiation.

Another development with subcontracting today is where the client enters into direct contract with a whole series of specialists, each responsible for packages under the direction of a management contractor. In this case there will also be a design element shared between the architect/engineer and the specialist, the pre-contract service being perhaps met by a fee to cover the impact of product or component development. It is necessary to point out the varied approaches to obtaining components and to the fact that the specification writing and the role of the specialist have changed in a fundamental way.

The other shift in emphasis is the concept of quality assurance which aims to indemnify the client against failure of the building components and their design. The process is covered in BS 5750 and is intended as a procedure to be adopted by each branch of the building industry — architect, engineer and surveyor managers, designers and specialists. The working arrangements in the industry have led to certification of certain firms, but in reality, a firm can comply with BS 5750 by showing intention to proceed in agreement with detailed procedures laid down in the Standard.

1.2 Requirements for components

The craft basis for describing components such as brick and brickwork or stone and stonework is best studied in British Standard Codes of Practice. It is worth referring to specific examples to see how complicated the additive methods of British Standards have become. In the case of brick, one has to look at BS 187 : Part 2, 1180, 3679, 3921 and 4729. A typical production specification for bricks gives a description of the materials, sizes (including tolerances), classification and appearance. Tables are given for physical or mechanical requirements with recommendations for specific uses of the bricks and the mortar mixes to be used. Mention is also made of the maker's certificates that can be called for. This is typical of a 'production specification' which prescribes matters in absolute terms without recourse to the expertise of the manufacturer.

By comparison, the 'performance specification' for a component is drawn up to provide exact data about requirements and function. By not being specific about production methods, the manufacturer is allowed to select materials and means of assembly; the incentive to develop economic methods of manufacture therefore rests with the specialist suppliers. They are expected to consider current and innovatory techniques of manufacture, the properties and behaviour of appropriate materials and to take into account user requirements as well as anthropometric and ergonomic data. The user requirements and related data are usually part of the target area devised by the specification writer. The aesthetic considerations imposed by the designer will relate to general arrangement layouts, profile and proportion, while the production drawings will be part of the package of information produced by the manufacturer to the approval of the designer. The discrete package of subcontracts used in tendering procedures today favours the 'performance' approach in order that the client obtains the most economic solution from specialist suppliers within the building industry. The designer's role falls into two parts: firstly, the initial development of performance criteria and, secondly, the ability to access and advise on the manufacturer's submission. Designing within this field is often a matter of architectural and engineering skill since elements like external cladding or stairs and balustrading perform a structural role.

1.3 Specification writing for components

The necessary references and technicalities for preparing specifications (whether craft or performance based) are set down in a useful guide by Christopher Willis titled *Specification Writing for Architects and Surveyors* published by Blackwell Scientific in 1991. The key subdivisions for a performance specification are as follows:

- Component description and use in general terms (this should enable an intended supplier to decide whether they are able to submit a price or to tender).
- Inclusion of CI/SfB classifications.
- Invitation to supplier concerning type and quality of component already manufactured by them and which they consider satisfies the specification.
- Reference to appropriate Codes of Practice, British

Standards and to governing clauses of the Building Regulations. (It is usual to detail tests needed under British Standards and Codes of Practice.)
- Specific data concerning layout and profile.
- Required maximum and minimum life for a component.
- The length and nature of guarantees needed and the nature of insurance cover. (It is common practice for the manufacturer to draw up the insurance policy and for the client to pay the premiums.)
- The manufacturer should be asked to provide details of the recommended method for maintenance. Clearly a planned sequence for the life of the component can be written into the insurance documents and be part of the documentation given to the client at handover.
- It is advisable to ask for written confirmation from the manufacturer regarding delivery period and their requirements on site for storage and attendance.

The success of performance specifications depends upon feedback to eradicate previous mistakes. The Building Research Establishment (BRE) and the Property Services Agency (PSA) are useful sources of information, a valuable guide from the early days of the new technique being a report *Performance Specification for Whole Buildings*, edited by Harrison for BRE Studies and covering 1974–82. Clearly an agreed list of terms or headings will provide a common means of communication between the building team, a useful source being the HMSO publication referenced DC 9 and titled *Performance Specification for Building Components*. Table 1.1 is taken from this and provides a summary of the main properties in 'check list' form for various components; qualifications regarding inclusion are set down in the index at the head of Table 1.1. The numbering system is based on the *Master List of Properties* (C1B Report No. 3, 1964) and gives a direct link between component and material properties.

1.4 Component testing and quality assurance

The following paragraphs set down the situation in 1993 where the British Standards Institution (BSI) and the British Board of Agrément still form the assessment service for new materials, products, components and services. At the time of writing, however, the new Eurocodes and Euronorms emanating from the EC are still largely in the drafting stage and have not become mandatory.[5] The discussion taking place between the member countries will produce an amalgam of building codes which one day will be enforceable in preference to national regulations or to standards like those published by the BSI. In an ideal Europe all the new codes would be in place at one time, but the likelihood in the UK will be a lengthy interregnum with components which cover many materials being covered by

Table 1.1 Component properties to be considered when preparing a performance specification with reference to a selected list of components

This is a check list and it will depend upon the particular requirements to be met whether a property should be specified or not. In certain cases the performance specification writer will be unable to set quantified values for the properties but may request the component manufacturer to furnish details in respect of a component offered in response to a performance specification.

Heading	CIB No.	Window	Roof finish	Partition	Internal door set	Ceiling	Floor finish
GENERAL INFORMATION	1.1						
Description of component	1.1.01	X	X	X	X	X	X
Type and quality		X	X	X	X	X	X
Identification of standards,	1.1.02						
quality mark	1.1.03	X	X	X	X	X	X
Purpose and use	1.1.04	X	X	X	X	X	X
Accessories	1.1.05	X	X	X	X	X	X
COMPOSITION and MANUFACTURE	1.2						
Composition	1.2.01	X	X	X	X	X	X
Manufacture and assembly	1.2.02	X	X	X	X	X	X
SHAPE, DIMENSION, WEIGHT	1.3						
Shape	1.3.01	X	X	X	X	X	X
Dimension	1.3.02	X	X	X	X	X	X
Geometric properties	1.3.03	X	X	X	X	X	X
Volume	1.3.04	—	—	—	—	—	—
Weight	1.3.05	X	X	X	X	X	X
GENERAL APPEARANCE	1.4						
Character of visible face	1.4.01						
Evenness	1.4.01.1	X	X	X	X	X	X
Appearance	1.4.01.2	X	X	X	X	X	X
Transparency, translucency	1.4.02	X	—	X	X	X	—
PHYSICAL, CHEMICAL AND BIOLOGICAL PROPERTIES	1.5						
Specific weight	1.5.01	X	X	X	—	X	X
Internal structure	1.5.02	X	X	X	X	X	X
Chemical formulation and							
material specification	1.5.03	X	X	X	X	X	X
Penetration of air and gases	1.5.04	X	X	X	X	X	X
Properties relating to the presence							
of water	1.5.05						
Moisture content	1.5.05.1	X	X	X	X	X	X
Solubility in water	1.5.05.2	X	X	X	X	X	X
Capillarity	1.5.05.3	X	X	X	X	X	X
Drying and evaporation	1.5.05.7	X	X	X	X	X	X
Moisture movement	1.5.05.8	X	X	X	X	X	X
Water absorption	1.5.05.4	X	X	X	X	X	X
Water penetration	1.5.05.5	X	X	X	X	X	X
Water vapour penetration	1.5.05.6	X	X	X	X	X	X
Thermal properties	1.5.06.1						
Thermal movement	1.5.06.1	X	X	X	X	X	X
Specific heat	1.5.06.2	—	X	X	—	X	X
Freezing and melting-point	1.5.06.3	—	—	—	—	—	—
Radiation coefficient	1.5.06.4	X	X	X	—	X	X
Thermal conductance	1.5.06.5	X	X	X	X	X	X
Warmth to touch	1.5.06.6	—	—	X	X	—	X
High and low temperatures	1.5.06.7	X	X	X	X	X	X
Thermal shock	1.5.06.8	X	X	X	X	—	X

Table 1.1 continued

Heading	CIB No.	Window	Roof finish	Partition	Internal door set	Ceiling	Floor finish
Strength properties	1.5.07						
Tension	1.5.07.1	X	X	X	X	X	X
Compression	1.5.07.2	X	X	X	X	X	X
Shear	1.5.07.3	X	X	X	X	X	X
Bending	1.5.07.4	X	X	X	X	X	X
Torsion	1.5.07.5	X	X	X	X	X	X
Impact	1.5.07.6	X	X	X	X	X	X
Hardness	1.5.07.7	X	X	X	X	X	X
Resistance to fatigue	1.5.07.8	X	X	X	X	X	X
Mechanical properties	1.5.08						
Resistance to mechanical water	1.5.08.1	X	X	X	X	X	X
Resistance to the insertion and extraction of nails and screws	1.5.08.2	X	X	X	X	X	X
Resistance to splitting	1.5.08.3	—	X	X	X	—	X
Resistance to tearing	1.5.08.4	—	—	X	—	—	X
Resistance to bursting	1.5.08.5	—	—	—	—	—	—
Rheological properties (flow and deformation)	1.5.09	X	X	X	X	X	X
Frictional resistance	1.5.10						
Coefficient of friction	1.5.10.1	—	X	—	—	—	X
Degree of slipperiness in use	1.5.10.2	—	X	—	—	—	X
Adhesion	1.5.11	—	X	—	—	—	X
Acoustic properties	1.5.12						
Sound absorption, sound reflection	1.5.12.1	X	—	X	X	X	X
Sound transmission	1.5.12.2	X	—	X	X	X	X
Optical properties	1.5.13						
Light absorption, light reflection	1.5.13.1	X	X	X	X	X	X
Light transmission	1.5.13.2	X	—	X	X	X	—
Light refraction and dispersion	1.5.13.3	X	—	X	X	X	—
Optical distortion	1.5.13.4	X	—	X	X	X	—
Electrical properties	1.5.14						
Electrical conductivity (electrical resistance)	1.5.14.1	X	X	X	X	X	X
Dielectric constant	1.5.14.2	—	—	—	—	—	—
Liability to develop and shed electrostatic charges	1.5.14.3	X	X	X	X	X	X
Effect of sunlight	1.5.15	X	X	X	X	X	X
Effect of electromagnetic and particle radiation	1.5.16	—	—	X	X	X	X
Effect of freezing conditions	1.5.17	X	X	X	X	X	X
Effect of fire	1.5.18						
Combustibility	1.5.18.1	X	X	X	X	X	X
Fire resistance	1.5.18.2	X	X	X	X	X	X
Surface spread of flame	1.5.18.3	X	X	X	X	X	—
Effect of chemicals	1.5.19	X	X	X	X	X	X
Effect of impurities	1.5.20	X	X	X	X	X	X
Effect of fungi, micro-organisms and insects	1.5.21	X	X	X	X	X	X
Effect of other building materials	1.5.22	X	X	X	X	X	X
Changes of behaviour during use	1.5.23	X	X	X	X	X	X
Setting time	1.5.23.1	—	X	—	—	X	X
Heat evolution in preparation and application	1.5.23.2	—	X	—	—	X	X
Change in volume	1.5.23.3	X	X	X	X	X	X

Continued overleaf

Table 1.1   continued

Heading	CIB No.	Window	Roof finish	Partition	Internal door set	Ceiling	Floor finish
Properties important from the point of view of hygiene	1.5.24						
Toxicity	1.5.24.1	X	X	X	X	X	X
Odour	1.5.24.2	X	X	X	X	X	X
Taintability	1.5.24.3	X	X	X	X	X	X
Tendency to deposit dust	1.5.24.4	—	X	X	—	X	X
Injury to skin	1.5.24.5	X	X	X	X	X	X
Liability to vermin infestation	1.5.24.6	X	X	X	X	X	X
Liability to become dirty, ease of cleaning	1.5.24.7	X	X	X	X	X	X
Safety	1.5.24.8	X	X	X	X	X	X
Tendency to deposit dust	1.5.24.4	—	X	X			
DURABILITY	1.6						
Durability of the component or assembly	1.6.01	X	X	X	X	X	X
Durability of specified component parts	1.6.02	X	X	X	X	X	X
Guarantee of durability	1.6.03	X	X	X	X	X	X
PROPERTIES OF THE WORKING PARTS, CONTROLS, ETC.	1.7						
Method of operation	1.7.01	X	—	X	X	—	—
Connection data	1.7.02						
Mechanical connection	1.7.02.1	X	—	X	X	—	—
Connection to power supply	1.7.02.2	X	—	X	X	—	—
Performance data	1.7.03						
Mechanical data	1.7.03.1	X	—	X	X	—	—
Capacity	1.7.03.2	—	—	—	—	—	—
Other performance data	1.7.03.3	X	—	X	X	—	—
Consumption of energy and ancillary materials	1.7.04						
Supplied energy	1.7.04.1	X	—	X	X	—	—
Ancillary materials	1.7.04.2	X	—	X	X	—	—
Efficiency	1.7.05	—	—	—	—	—	—
Manoeuvrability and control	1.7.06	X	—	X	X	—	—
Other technical data	1.7.07						
Mechanical	1.7.07.1	—	—	—	—	—	—
Thermal	1.7.07.2	—	—	—	—	—	—
Electrical	1.7.07.3	—	—	—	—	—	—
Secondary effects and disturbances during operation	1.7.08	X	—	X	X	—	—
WORKING CHARACTERISTICS	1.8						
Ease of handling	1.8.01	X	X	X	X	X	X
Consistence, workability, working time	1.8.02	—	—	—	—	—	—
Ease of cutting, sawing, bending, etc.	1.8.03	—	X	X	X	X	X
Capability of being jointed to other components	1.8.04	—	X	X	X	X	X
Fixing	1.8.05	X	X	X	X	X	X
Surface treatments	1.8.06	X	X	X	X	X	X
Capability of withstanding rough handling	1.8.07	X	X	X	X	X	X
Capability of withstanding storage	1.8.08	X	X	X	X	X	X

a mixture of BSI and Eurocodes a time surely when building insurance will have higher premiums.

The deadline of 1 January 1993 has passed with barely 1 per cent of the proposed European Standards in place. The BSI are modifying their standards to meet EC requirements, but to date the components have been of minor interest such as outlets from shower trays. The new terminology refers to 'Euronorms' for individual items while 'Eurocodes' cover structural matters. It would appear that the change-over will take the rest of this decade.

The British Board of Agrément plays a crucial role in this period and is the focus of European technical approval in the UK. It provides the superintending role for laboratory testing for components, materials, products and building processes. The resources of the BRE are also involved to assist the Board of Agrément with their assessment service based on examination, tests and detailed investigation. The aim is to provide the best technical opinion possible within the knowledge available. Hitherto, before European involvement, components or processes already covered by a British Standard did not fall within the scope of the Board of Agrément. That aspect has changed totally with the hopeful deadline of 1 January 1993 for establishing Eurocodes and Euronorms. In times past Agrément certificates operated on a different basis and were renewable after 3 years and which allowed for failures to be monitored and for retrospective measures to be described for manufacture and application. It is worth studying Agrément archives in detail, particularly where innovative techniques have been developed.

The testing services are today performed by private laboratories supervised through the Board of Agrément or else the work is carried out in consultation with the Building and Engineering Section of the BSI. The intent is a testing service for a wide range of building materials and components, including windows and doors, wall claddings, ironmongery, and plumbing units to arrive at new European Standards for Eurocodes and Euronorms. Testing facilities enable the following forms of tests to be carried out: weathertightness and simulated wind gust loading; mechanical strength and related characteristics of metallic and plastics components and materials; impact, pressure, endurance and corrosion; artificial weathering; and assessment of finishes on components or materials. The other ongoing purpose of the service is to assist manufacturers in the development and marketing of the products, and to provide building designers and specifiers with data on performance of products being considered for specific building projects. It also assists trade associations, government departments and manufacturers in the development of performance-based product specifications. Items are tested against national and international performance standards, government department specifications, manu-

Figure 1.1 British Standard kitemark and quality assurance registered firm symbol.

facturers' or users' own specific requirements.

Under the BSI system, a material or component which completes a satisfactory test receives a *BSI Test Report* as well as a *kitemark*. The latter is a registered trade mark owned by BSI and may only be used by manufacturers licensed by them. The use of this symbol (Fig. 1.1) on a product indicates satisfactory independent testing by the BSI of product samples against every aspect of the appropriate British Standard. A further test sequence can occur where the product is concerned with safety (for example, multiply security glazing) and where the kitemark would carry an additional 'safety mark'. The kitemark will be replaced under Common Market legislation by a CE marking (the European Communities' conformity mark), but only related to individual products.

The topic of 'quality assurance' has come to the fore in the 1980s and relates back to BS 5750 : 1979 titled *Quality Systems*. These are assessed by BSI as part as their *Quality Assurance Services' Certification and Assessment Scheme* which sets out the organization, responsibilities, procedures and methods involved in manufacturing the product. Where a firm's 'design capability' is to be included in the assessment, BS 5750 : Part 1 is used. Of great significance in an increasingly competitive world for consultancy organizations pursuing fewer project commissions, firms offering professional design services — such as architects and engineers — are also seeking certification under this scheme. Otherwise, where a firm works to a published specification or the customer's specification, BS 5750 : Part 2 is used.

Firms registering for the scheme are required to have a documented quality system which complies with the appropriate Part of BS 5750 and a related *quality assessment schedule*. Before an assessment is arranged, a detailed appraisal of the applicant's documentation for compliance with these requirements is undertaken. The applicant is then notified of any significant omissions or deviations from the requirements in order that suitable amendments can be made prior to the assessment. Once the applicant's documented procedures are considered to be satisfactory, an assessment visit is arranged.

The assessment itself involves an in-depth appraisal of the firm and a requirement to demonstrate the practical application of the documented procedures. Where an assessor discovers a deviation from the requirements or witnesses a non-compliance with the documented procedures, a *discrepancy report* is given. After any deviations have been answered or rectified, or if no deviations are discovered, initial registration is granted and a *certificate of registration* can be issued. The registration assessment programme continues subsequent to granting the initial registration to confirm that the quality system is operating effectively throughout the lifetime of individual projects falling within the scope of registration. Post-registration assessment visits are carried out at pre-agreed intervals, normally at the rate of two visits within the first 6 months and a minimum of two visits per annum thereafter. Any later and acceptable alterations to the details shown on the certificate of registration entitles the firm to use the *registered firm symbol* (Fig. 1.1) on letterheadings, in advertisements and for other promotional purposes relating to the *organization and management* of the firm, not for products, but registered firms now include contractors, manufacturers and professionals like architects, engineers and surveyors. Risks from bankruptcy amongst building professionals are aspects to be considered in the 1990s.

In conclusion, there appears at present to be very limited correlation with Eurocodes so that consideration of quality assurance under the BSI is a matter for work within the UK under BS 5750. The international equivalent is ISO 9000.

1.5 Methods of manufacture

It is necessary for the designer of a component to understand the discipline of factory production so that collaboration with the manufacturer at design stage will be effective. Two methods of factory work are appropriate to the making of building components, either flow line or batch production.

Where *flow line production* is in operation, a stream of component parts in various stages of completion travel by conveyor belt through a number of work positions to completion. At each position one or more operations are carried out until at the final position, at the end of the assembly line, the component is complete. The operations are standardized at each work position and careful organization is required so that the necessary materials are always available. Flow line production methods can be highly automated, which means a high level of capital investment in automatic machinery with a minimum of labour. With this type of manufacturing process it is not easy for alternative operations to be carried out at a particular work position. The situation is often resolved by employing a secondary production line where, for example, differing forms of opening light are added to standard

window frames while plain frames (without lights) proceed to completion.

However, a flow line system is most efficient when uninterrupted by alternative operations. This presupposes complete standardization, which in its turn requires the development of a co-ordinated system of sizes and dimensions. Flush doors are normally produced by a flow line system. It is obvious from this example that where a large number of components are involved and the quality of the raw material can be carefully controlled, then factory production and factory-applied finishes produce an article of better quality and value than traditional methods on site.

Batch production involves the setting up of machinery to manufacture a batch of components. An example of this would be the moulding of jambs and rails for a timber window, followed by the manufacture of a batch of sill sections after the machinery has been reset. Batch production is not only adopted for machinery operations. It is also used for the assembly of parts where quantities, in mass production terms, are comparatively small, or where there is too much variety to allow the efficient use of flow line techniques (see Figs 5.31 and 5.32).

Batch production is also the appropriate method where the cost of processing is low in comparison to the cost of setting up the process. In connection with the production of timber components, woodworking machines have a high rate of production in relation to the time taken to set the machine up. The relationship of the cost of production to the cost of setting up can, however, be made more economical if the variety of sections and mouldings can be reduced for a particular component. Thus, where the same moulded cross-section of timber can be used for various parts of, say, a glazed window wall, then the cost of one machine setting will be spread over the cost of the total length of the moulded section for the job.

In practice, combinations of both flow line and batch methods are used. For example, in the factory production of timber components the machining will probably be done by batch production and the assembly of the timber sections into the completed component will be carried out by flow line techniques. However, it must always be remembered that production techniques are continuously being examined, modified and improved in an effort to overcome difficulties and disadvantages. It is essential that the consultation at design stage should be fully developed to facilitate this process.

A single production line would be a single and continuous operation with the input of raw materials at one end and the output of finished goods at the other. At the start of production line there will be a store of raw materials which will allow certain fluctuations in delivery. Where one operation takes longer to perform than the others the line will have to be split, or additional machines and/or labour introduced. Storage will also have to be provided at the end of the production line to absorb fluctuations in demand. The theoretical layout of a production line will almost certainly be inhibited by physical limitations of factory space.

The proportion of overheads is not so high with mass production when compared with the production of a small number of components on a small scale. The percentage of the working year during which the factory is operating to full capacity is also a significant consideration in the cost per unit of the component. Machinery should be capable of being modified so that improvements in the design of a component can be incorporated without undue capital expenditure.

In order to produce components economically there is an optimum output in terms of the number of components produced relative to the nature of component and type of plant used. However standardized a production system is, it is inevitable that some components will be required to be non-standard or may be required in such small numbers as to make them uneconomic to produce on the standard production line. The higher the degree of automation the more difficult it is to produce non-standard items, and it must be expected that they will be more expensive and have an extended delivery period. It follows from this that manufacturing techniques which can produce related components economically over the widest possible range will be more acceptable in the long term.

Other industries have embarked upon 'flexible manufacturing systems' (FMS), particularly in car production where vehicles are assembled by a team of workers, each having flexible and multiple roles to undertake instead of the compartmental approach of the traditional assembly line. A roofing manufacturer of felt and sheet coverings has developed a machine which can produce a whole range of products to differing specifications on one set of machinery. This application of FMS gives the maker the facility to respond to the market without the batch ordering that was common when production lines had to run for specific periods to cover the setting up and tooling costs.

To allow a manufacturer greater control over the production and detailing it is good practice to invite quotations for components on the basis of a performance specification which indicates the parameters within which the product must perform. The successful tenderer at this stage can then be consulted in respect of detail and development work.

The various processes in the factory production of a range of standard timber windows are detailed in Chapter 5 and for standard timber doors in Chapter 6. It is very important that production can proceed efficiently without delays caused by imprecise information from the designer or management.

1.6 Implications of computer programs

Numerous *computer programs* are now available which ensure that precise details are made available for the efficient production of assemblies such as cupboard joinery, staircases, stud frames and trusses. The software controls materials, quantities, dimensions and profile needed for each component. It can also manage cutting schedules and give drawn or written directions as to assemblies made at work stations apart from ensuring delivery of materials to each workplace. The design package includes full size drawings for profiles together with elevational, plan and section layouts. Stock-taking is another aspect together with estimating and time for production. The programs are extendable to whole building frames in timber, cold-formed steel, structural sections and precast concrete. Timber house shells have been produced under computer programs for many years; experience has shown that computer aids to the assembly line increase house production in the factory by 40 to 50 per cent. Computer-aided manufacture (CAM) as outlined for carpentry (trusses), joinery or house framing can be extended into the design field by computer-aided design (CAD). The obvious applications are calculations by the varying geometry of trusses, but more sophisticated designing is employed in the development of complex components such as rain-screen façade elements or staircases. In the latter example, many manufacturers are offering a substantial design service which embraces the total production of design drawing and the requisite details for production work. This aspect is in parallel with the changing roles within 'design and build' contracting, where design impact is increasingly the responsibility of the manufacturers or specialist fabricators, be they makers of components, façade claddings, joinery or staircases.

1.7 Dimensional co-ordination

The traditional pattern of trade following trade with materials cut to fit together on site made it possible for later trades to make good any inaccuracies in completed work. Rationalization of this building process involves the use of an increasing range of factory-produced components which avoids the waste arising from cutting on site. For this to be the most effective, it is essential that the component dimensions are co-ordinated by reference to an agreed range of sizes.

In accepting agreed dimensional standards for components, the use of non-standard or *purpose-made* components is largely avoided, thereby assisting design decisions and speeding the production of working details. Manufacturers can make more effective use of their production resources (materials, labour and plant) and on-site processes can be speeded up because of familiarity with the components and erection permutations. However, it is important that manufacturers co-operate with the real world of building and accept a dimensional framework for their components that not only recognizes the process by which they are made, but also an industry-wide discipline. In essence, this means that special consideration must be given by all parties concerned to the joints and jointing methods between the components. This is the point where, on a building site, production methods meet realities of construction techniques.

Dimensional co-ordination is a system of arranging the dimensional framework of a building so that standardized components can be used within the framework in an interrelated pattern of sizes. In this way it can be closely connected with the overall development of building technology and the evolution of new building processes.

1.8 References and terminology for dimensional/modular co-ordination

The most relevant British Standards which give detailed information on dimensional and modular co-ordination are: given as follows. A word of warning needs to be given that not all the terms are as yet common parlance in the building industry. See Figs 1.2 and 1.3 and Table 1.2 with reference to keywords.

BS 4606 : 1970 *Recommendations for the co-ordination of dimensions in building*. Co-ordinating sizes for rigid flat sheet material used in building.

BS 4643 : 1970 *Glossary of terms relating to joints and jointing in building*.

BS 5568 *Building construction* Part 2 : 1978 *Modular co-ordination*. Specification for co-ordinating dimensions for stairs and stair openings.

BS 5606 : 1987 *Accuracy in building*.

BS 6100 *Glossary of building and civil engineering terms* Part 1 *General and miscellaneous* Section 1.5 Operations: associated plant and equipment. Subsection 1.5.1 : 1984 Co-ordination of dimensions, tolerances and accuracy.

BS 6222 *Domestic kitchen equipment* Part 1 : 1982 *Specification for co-ordinating dimensions*.

BS 6750 : 1986 *Specification for modular co-ordination in building*.

PD 6446 : 1970 *Recommendations for the co-ordination of dimensions in building*. Combinations of sizes.

DD 22 : 1972 *Recommendations for the co-ordination of dimensions in building*. Tolerances and fits for building. The calculation of work sizes and joint clearances for building components.

An understanding of the various terms used in connection with dimensional co-ordination in building is necessary, since descriptions which previously have been used loosely now have specific meanings. The following definitions have been taken from BS 6100 *Glossary of building and civil engineering terms* Part 1 *General and miscellaneous*. Subsection 1.5.1 : 1984 Co-ordination of dimensions, tolerances and accuracy:

Building element Major functional part of a building, e.g. foundation, floor, wall services.

Component Product manufactured as a distinct unit to serve a specific function or functions.

Assembly A set of building components used together.

Dimensional co-ordination Convention on related sizes for the co-ordinating dimensions of components and the structures incorporating them, for their design, manufacture and assembly.

Module Unit size as an incremental step in dimensional co-ordination.

Basic module Fundamental module, the size of which is selected for general application to building and components. Its value has been standardized as 100 mm.

Modular size Size that is a multiple of the basic module (100 mm).

Modular component Component whose co-ordinating sizes are modular.

Multimodule Module whose size is a selected multiple of basic module (100 mm).

Submodular increment Increment of size the value of which is a selected fraction of the basic module (100 mm).

Modular co-ordination Dimensional co-ordination employing the basic module (100 mm) or a multimodule.

Accuracy Quantitive measure of the magnitude of error.

Tolerance Permissible variation of the specific value of quantity.

Grid Rectangular co-ordinating reference system.

Modular space grid Three-dimensional grid in which the distance between consecutive planes is the basic module or multimodule. This multimodule may differ for each of the three dimensions of the grid.

Modular line Line formed by the intersection of two modular planes.

Modular plane Plane in a modular space grid.

Planning module Multimodule adopted for specific applications.

Preferred modular size Modular size or multimodular size which is selected in preference to others.

Work size Target size of a building component specified for its manufacture.

Reference system A system of points, lines and planes to

which the sizes and positions of a component, assembly or building element may be related.

Reference space   Space assigned in a structure to receive a component, assembly or building element including, where appropriate, allowances for tolerances and joint clearances. The space is bounded by reference planes which are not necessarily modular.

Co-ordinating plane   Plane by references to which one component is co-ordinated with another.

Co-ordinating space   Space bounded by co-ordinating planes, allocated to a component including allowances for tolerances and joint clearances.

Co-ordinating dimension   A dimension of co-ordinating space.

Co-ordinating size   The size of a co-ordinated dimension.

Zone   Modular or non-modular space between modular, planes, which is provided for a component or group of components which do not necessarily fill the space, or which may be left empty.

1.9 Modular co-ordination

It is necessary to establish a rectangular three-dimensional grid of basic modules into which the component will fit — see Fig. 1.2. A *module* is the basic dimension of a unit from which all other measurements can be derived. The term 'module' comes from the Latin *modulus* (small measure) and has been used in building ever since the first century BC.

Notwithstanding the comments made in section 1.7 concerning the acceptability of standardized component sizes between manufacturer and the actualities of the building site, there is now an *internationally agreed basic modular dimension of 100 mm*. This is used to give

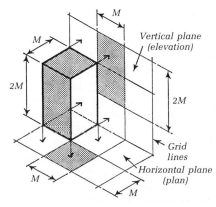

Figure 1.2   Three-dimensional grid of basic modules.

guidance to designers and manufacturers about the fundamental unit of size for the horizontal and vertical co-ordinating dimensions of buildings, their spaces, components and materials. This basic module is denoted by the letter M in the accompanying text.[6]

The precise overall modular dimension for a particular component will depend upon the function of the component as well as the materials from which it is made, the manufacturing processes involved, the most appropriate method of transportation from the factory to site and the ease of storage and/or subsequent positioning in a building. BS 6750 : 1986 recommends that components should be manufactured in *multimodular increments* of 3M, 6M, 12M, 15M, 30M and 60M; and in *submodular increments* (fractions of the basic module) of either 50 mm as first preference or 25 mm as second preference — see Table 1.2. The adaptation of the basic module in this way permits internationally supplied components to be used with those from the UK. Modular sizes do not give the actual or working size of the component but only allots space for it in a building.

Modular components are often required to be used in conjunction with components which are non-modular. An example of this is given in Fig. 1.3 taken from BS 6750. This indicates the use of modular windows with British Standard size bricks, i.e. 65 mm high with 10 mm joint. Used in an appropriate number of courses, overall modular sizes can be obtained for the brickwork or for a combination of window and brickwork. Alternatively, as in Fig. 1.3 (*d*) and (*e*) a filler section can be used to obtain overall modular co-ordination. The brick industry's attempt at producing modular bricks (200 and 300 × 100 or 75 as well as 200 × 100 × 75) did not meet with great success, perhaps because of a combination of the cost needed for new moulding equipment and the resulting break from the traditional appearance of the brickwork. In practice they were as economic in labour costs as conventional sized brickwork.

1.10 Co-ordinating planes and zones

The use of modular grids for the setting out of the spaces occupied by components is described in BS 6750 : 1986. This British Standard makes use of vertical and horizontal *co-ordinating dimensions* to define a regular framework within which to design and to which components and assemblies may be related. The dimensions define controlling planes and are shown by the graphics as in Fig. 1.4. For appropriate modular sizes of these planes reference should be made to Table 1.3(a) and (b).

Co-ordinating planes represent the key reference lines for setting out the building, indicating load-bearing walls

Table 1.2 Preferred modular sizes based upon the multimodules for horizontal and vertical dimensions (BS 6750 : 1986 Table 1)

Multi-modules	Preferred modular sizes in multiples of M																											
3M	3	6	9	12	15	18	21	24	27	30	33	36	39	42	45	48												
6M		6		12		18		24		30		36		42		48	54	60	66	72		78	84	90	96			
12M				12				24				36				48		60		72			84		96		108	120
15M					15					30					45			60			75			90		105		120
30M										30								60						90				120
60M																		60										120

Notes:
1. The preferred modular sizes that are selected in preference to other sizes for horizontal and vertical co-ordinating dimensions are primarily intended for the sizing of components, groups of components and spaces.
2. The sizes derived from 3M and 6M have been restricted in the table to the limits shown.
3. The 15M, 30M and 60M series correspond to the series in a system of preferred numbers which contain the factor five. These series can also be extended to use larger increments in the series such as 120M or larger.
4. In the selection of sizes from the table, preference should be given to the series of the largest multimodule compatible with functional requirements and economic design.

and columns, storey heights and other boundaries as shown in Fig. 1.5(a) and (b); other modular grid lines are used between controlling planes for the location of secondary components. Zones are located between vertical or horizontal co-ordinating planes where it is desirable to allocate a modular space for specific functions, such as for floors and roofs. These zones contain the structure as well as the finishes, the services, the suspended ceiling and, where appropriate, allowances for camber, fall and deflection. Similarly, zones for walls, partitions and columns contain their structure and finishes.

Figure 1.5 also indicates two methods of locating horizontal co-ordinating planes in relation to loadbearing walls and columns: on the axial lines of loadbearing walls or columns (Fig. 1.5(a)) or on the boundaries of the zones (Fig. 1.5(b)). Zones for columns and loadbearing walls should be selected from the following range: 100, 200, 300, 400, 500 and 600 mm. If greater widths are required they should be in multiples of 300 as first preference, or 100 as a second preference.

It is obviously important that grid lines for co-ordinating planes, zones and the positioning of components should always be shown on working drawings so that the builder can then set out the profiles on site without ambiguity. The phenomenon of one component infringing on the space which should be occupied by its neighbour and this effect becoming cumulative, is known as creep, and is avoided by the use of face grid lines (Fig. 1.7).

A range of modular co-ordinating dimensions and zones were made applicable to public housing in the late 1970s, which retain relevance in terms of floor-to-floor heights and the use of staircases (Table 1.4).

1.11 Working sizes of components

The basic size of a component is bounded by the modular grid (either between co-ordinating planes or a modular part thereof). From this basic size the *working size* of the component can be established as indicated in Fig 1.7(a) and (b).

In order to enable components to fit together without the need for cutting down to size on site or using excessively wide joints and cover strips to make up undersized units, it is necessary that they are manufactured so that their maximum and minimum sizes do not fall outside predetermined limits. As indicated in Fig. 1.7(a), for a component to fit correctly within its allocated modular grid it will always need to be slightly smaller than its basic size. The theoretical basic size established for a component, useful for planning and design purposes, will have to be reduced to achieve a working size to take into account inherent and induced deviations between specified and actual size as depicted in Fig. 1.7(b). The details in Fig. 1.8 illustrate the way in which current ideas for working sizes apply to standard door sets with dimensions given for components and gaps between items.

It is useful to look at graphic aids to design, a key reference being a BRE publication titled *Graphical Aids for Tolerance and Fits*. A handbook for manufacturers, designers and builders, it is helpful because it removes the need for complex and repetitive calculations and hence minimizes the time and effort at the design stage in considering all factors affecting jointing and fit of components. The vexed problem of tolerances and joints between components needs an introduction.

The shapes and sizes of elements within the assemblage of a building vary in their manufacture and are further altered by ageing and by creep or deflection due to performance underload. The latter engineering considerations are factors allowed for in the engineering design, but they have a bearing on the tolerances between structure and the enclosing skin of a building.

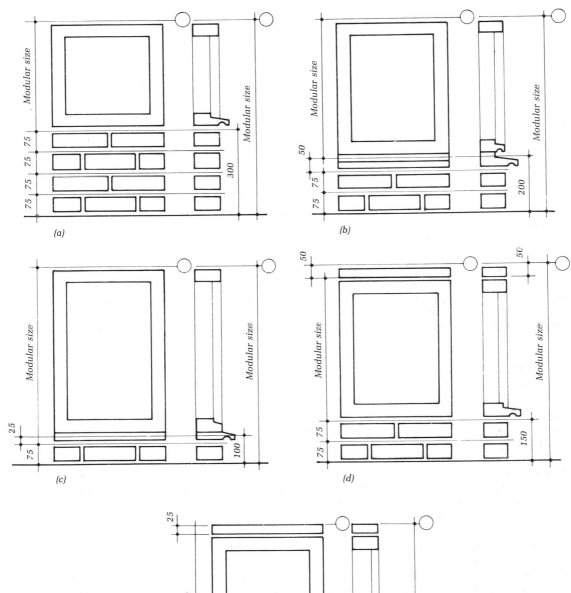

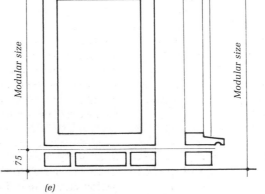

Figure 1.3 Brick coursing and modular sized windows (BS 6750 : 1986).

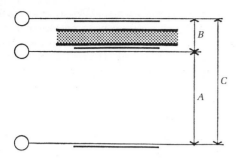

Figure 1.4 Floor to ceiling heights.

Table 1.3(a) Modular sizes for horizontal co-ordinating dimensions of spaces (*Source*: BS 6750 : 1986 Table 2)

Dimension/space	Range of space (mm)	Multiples of size
Zones for columns and loadbearing walls	100–600	3M or 1M
Centres of columns and wall zones	From 900	3M or 1M
Spaces between column and wall zones	From 600	3M or 1M
Openings in walls (e.g. for windows and door sets)	From 600	3M or 1M

Note: The first preference for the multiple of size in each case is 3M.

Table 1.3(b) Modular sizes for vertical co-ordinating dimensions of spaces (*Source*: BS 6750 : 1986 Table 3)

Dimension/space	Range of space (mm)	Multiples of size
Floor to ceiling and floor to floor (and roof)	Up to 3600	1M
	From 3600 to 4800	3M
	Above 4800	6M
Zones for floors and roofs	100–600	1M
	Above 600	3M
Changes of floor and roof levels	300–2400	3M
	Above 2400	6M
Openings in walls (e.g. for windows, including sills and/or subsills, and for door sets)	300 to 3000	3M or 1M

Notes:
1. For application of 75 mm sizes for bricks and 200 mm sizes for blocks.
2. Where the option of 3M or 1M is given, the first preference for the multiple of size is 3M.

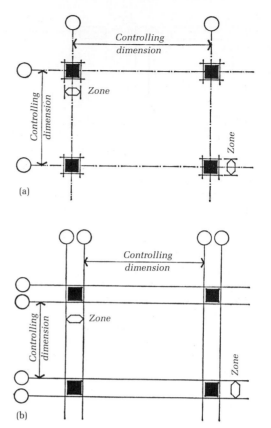

Figure 1.5 Horizontal controlling dimensions: (a) by axial lines, (b) by zoned boundaries.

There are *inherent deviations* due to the following physical characteristics of materials:

Shrinkage of concrete products.
Shrinkage, expansion and warping of timber.
Expansion and deflection of metal, brick and plastics.

There are additional *induced deviations* arising from manufacturing of work on site:

Manufacturing that gives rise to discrepancies in overall sizes and or flatness of surface or where deformation affects three-dimensional integrity.

Setting-out processes on site which result in inaccuracy of the prescribed guidelines or grids for locating the components.

Erecting and positioning the component against guidelines or grids so that slight misplacement occurs in height, length, verticality and/or differing angular qualities.

Fixing or seals between components, where these do not form an integral part of the component design. Typical problems occur where deviations in space available prevent fixings being inserted or for seals to be effective weatherings.

1.12 Tolerance and fit

Tolerances and the problem of 'fit' between varying elements of buildings are one of the most vexing problems. The whole situation is made more exasperating by the aesthetic endeavour for a smooth skin to buildings both externally and internally. Under the discipline of traditional construction regime architraves and/or cover battens would suffice without the complexities revealed in the latest standard details for fibre cement cladding or for door sets without joinery trim.

British Standards have grappled with this problem and often failed. BS 3626 : 1968 had to be withdrawn since the complex calculations erred so far in terms of inaccuracy that gap filling or gap covers between components became totally unrealistic. Today, tolerances and fit for building elements are covered by BS 5606 : 1990. This edition is a rewrite of earlier material from 1978 and devised to make for easier application. There are significant alterations to the initial attempts at deriving acceptable standards of accuracy. The following items are drawn from notes prepared by Julian Ryder-Richardson and Michael McEvoy for the Steel Construction Institute in 1990–91.

- *Characteristic accuracy* This is the key to the approach of British Standards and where attempts are made to establish tolerances within a defined statistical range. Sample inaccuracies were measured and initially based upon the assumption that, in 369 out of 370 cases, the permissible degree of inaccuracy would be representative of general site practice and termed 'characteristic accuracy'. The slackness on the 1978 assessment has been tightened with the statistical basis shifted to 21 cases out of 22. As previously, the British Standard defines two distinct causes of inaccuracy in building, 'induced or inherent' deviations. See section 1.11 for full explanation.

- *Tolerances* The method for determining tolerances has been simplified. Now, the locations of elements of structure are to be related to reference lines and levels. The tolerances anticipated for secondary elements (e.g. cladding, ceilings or partitions) are to be related to adjacent parts of the structure. In other words, the key setting-out zones are 'either side' dimension grids that encompass structure instead of the 'centre-line' grid (see Fig. 1.5(b)) which occurs at a greater remove from component joints. Under the 1990 edition of BS 5606 it will be necessary to establish a rigorous system of

monitoring site and suppliers' dimensions as well as specifying in tender documents the method of measurement to be adopted. If it appears that such systems cannot be imposed in practice it may be better to accept inaccuracies and design accordingly. It should always be remembered that higher standards of accuracy can be specified and obtained if more time is allocated to site management although the penalty is higher cost.

The following points are offered in conclusion to the discussion on tolerances (the designer needs to bear in mind that general factors are not covered by British Standards).

- *Aesthetics* This involves the visual quality of the building and where alignment of external components may be of the highest importance. By comparison, ceiling and floor lines may be critical internally while

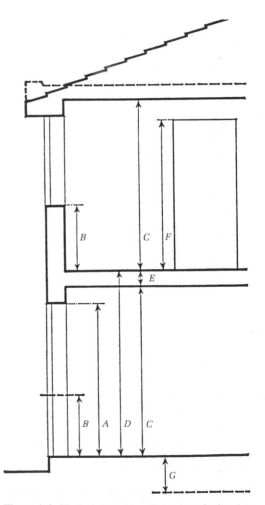

Figure 1.6 Vertical controlling dimensions for housing.

Table 1.4 Vertical controlling dimensions for housing

Measurement	Controlling dimension (mm)		
Window head height	2300 I		
	2100 II		
Window sill height	0	800	1200
	200	900	1400
	600	1000	1800
	700	1100	2100
Floor-to-ceiling height	2500	2300	
	2400	2100 — garages	
	2350	only	
Floor-to-floor height	2600		
	2700 — mandatory for public		
	sector housing		
Floor thickness	200	250	300
Door set height	2100		
Change of level	300	1300	1800
	600	1400	2000
	900	1500	2100
	1200	1700	2300
			2400

gaps between partitions and wall elements can be masked by flash gaps or cover plates.

- *Structural tolerances* There is a case to be made for greater accuracy in framing dimensions in order that a tighter discipline exists for structural connections. Reference should be to BS 5950. The questions posed by concrete could be: 'How inaccurate will it be?', while for steelwork which has the advantage of greater predictability, the question asked is: 'How accurate can it be?'.
- *Legal requirements* A particularly accurate relationship to site boundaries may be necessary or the maximum height of the building may be a requirement of statutory approvals. In the latter case the tolerance allowed in the camber of long-span beams for example, may become a significant issue.[7]
- *Practical requirements* It may be more cost-effective to require greater structural accuracy enabling fitting out (floor and ceiling systems) or cladding with industrialized non-structural cladding to be achieved to fine limits, instead of modifying the design components.
- *Fast track construction* Building at speed is inherently inaccurate, but casing out the frame allowing generous clearances for the structure may significantly reduce the lettable area of an office building. The simplicity of detailing often allocated at column junctions for 'fast track' may be at odds with the commercial determinants of achieving the maximum floor area.

The changes in procurement of buildings are outlined in the introduction to this chapter and have altered the role of designers and specifiers within the building industry.

They will also have the effect of reducing the architectural and engineering control over the design and fit of components unless those professions are willing to embrace the new scenario.

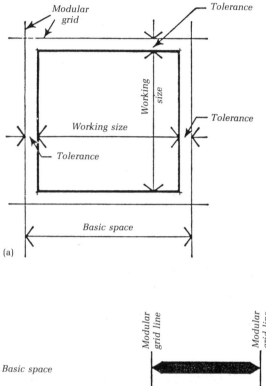

(a)

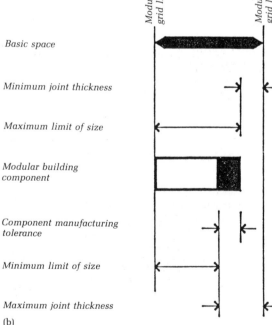

(b)

Figure 1.7 (a) Space relationship to grid; (b) relationship of a modular building to a planned grid.

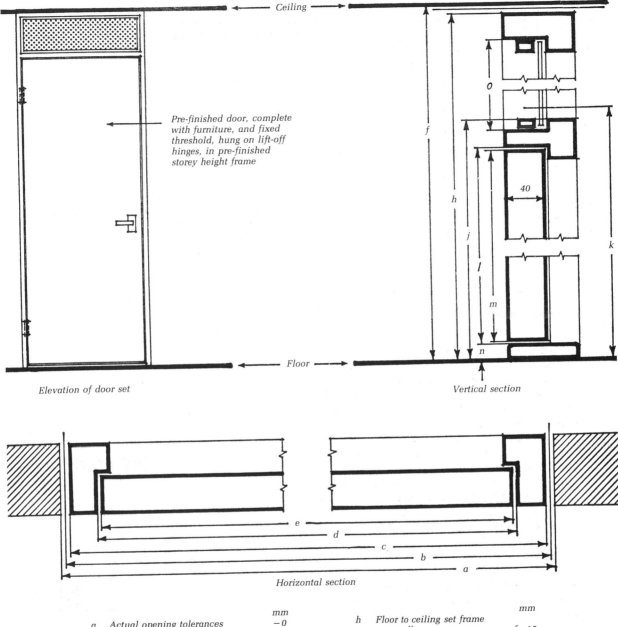

Elevation of door set

Vertical section

Horizontal section

		mm			mm
a	Actual opening tolerances (overall)	−0 +10	h	Floor to ceiling set frame overall	f−15
b	Nominal opening co-ordinating plane	900	j	Door height set frame overall	2090
c	Frame overall	890	k	Door height set opening	2100
d	Width between rebates	830	l	Door and clearances	2045
e	Door width fitted	826	m	Door height fitted	2040
f	Floor to nominal ceiling for housing:	2300 2350 2400	n	Threshold thickness	15
			o	Over panel rebate for housing:	177 227 277

Dimension f and b are grid line (or basic space) dimensions

Figure 1.8 Dimensions for standard door set.

Notes

1   Harold King and Alan Everett, *Components and Finishes*, Mitchell's Building Construction, B. T. Batsford, 1971. Harold King's chapter on component design is a real period piece on the optimistic role seen for industrialized building systems despite the lack of confidence that followed the Ronan Point collapse of 1968. For a more pessimistic viewpoint concerning industrialized buildings refer to writings by Alan Webb on Ronan Point or to Geoff Scott's book titled *Building Disasters and Failures* published by Construction Press Ltd in 1976.

2   Hermione Hobhouse, *Thomas Cubitt, Master Builder* by Macmillan, 1971. See chapters describing the Pimlico workshops and the building of Belgravia.

3   Alan Blanc, Michael McEvoy and Roger Plank, *Architecture and Construction in Steel* by Spon, 1993. See section 31.3 in the history of patent glazing components and Chapter 23 for notes on tolerances.

4   Archigram, a loose-knit group of radical British architects, including Warren Chalk, Peter Cook, Dennis Crompton and Ron Herron. The group published a journal *Archigram* from 1961 to 1970, typical designs employing large-scale components were Plug in City (1964−66) and Instant City (1969−70)

5   The situation at the time of writing is that the following Eurocodes have been published: EC 2 *Concrete*, EC 3 *Steel* with EC 4 *Composite Construction* due some time in 1993. At present mandatory use will only occur where public works are put out to tender to EC contractors. Use for general building work will probably be delayed by 7−10 years.

6   The basic module termed M is part of the language developed by enthusiasts for modular co-ordination. It is useful in terms of understanding the history of modular design but has not entered the common language of the construction industry.

7   The notion concerning beam cambers may seem far-fetched, but sites within the City of London are subject to stringent height and set-back limitations. Cambers for long-span beams affect clear heights between raised floors and suspended ceilings and, by aggregation, the total height of the building profile. It is often the reason for electing a framing design with column grids and modest framing dimensions, in order to reduce 'tolerances' in the vertical plane.

2 Demountable partitions

2.1 Introduction

The difference between demountable partitions and other lightweight construction in metal or timber stud is the fact that the components can be taken down and relocated without wastage of material and without materially affecting the surrounding structure. Partition elements that can be moved around at will are increasingly seen as part of office furniture and intregated with work stations; they are also seen as part of the environmental envelope to working areas with many patterns that form a three-dimensional modular system with raised service floors and suspended ceiling construction. This is particularly true in commercial buildings where cabling and electronic communication form a significant part of the activity; it is also applicable to advanced research establishments and to higher education. Modern industrial installations also rely on space modules that need endless alterations.

There are, however, locations where more traditional methods may be just as economic in use and particularly where a resident labour force is at hand for minor rebuilding work. Simple partition blocks or timber stud partitions often suffice for temporary partitions. On the other hand, metal stud and plasterboard partitions are speedier and can be finished in dry materials like plasterboard. They are also reusable, although the plasterboard will have in part to be renewed. Many clients prefer the latter method since fire and sound resistance can be obtained at low cost as compared with the high first costs and the skilled labour involved with sophisticated relocatable designs.

In essence, the truly demountable partition involves factory-made elements that frame together to form a building system with panels, frames and mechanical coupling. Finishes are factory applied except perhaps for sheet vinyl finishing and glazing. The construction is described in broad terms by British Standards (revised with BS 5234 : 1992 in Parts 1 and 2), but tenders can be invited on the basis of performance specification with required ratings in terms of stability, fire resistance and/or sound deadening.

The essential variations in form are set down in section 2.2. It should be borne in mind that many specialists simply put together assemblies from 'off-the-peg' components, hence the similarity in styling. There are, however, innovative manufacturers who will undertake prototype work with the development of specific profiles and trim, provided there is sufficient volume of repetition in the contract.

Movable partitions may be the answer where flexible space requirements demand the ability to extend areas or to rapidly subdivide. Mechanical operation will further speed matters. Placing vertical sliding or folding shutter doors where the components can be lifted overhead when not in use will be less restrictive of floor space than horizontal sliding or sliding folding 'concertina' doors which require spacious parking bays. Wicket gates (namely the individual door within a larger door panel) are often essential for means of escape apart from the practical aspects of access without 'motorizing' a whole movable partition to open up a wall.

Multipurpose spaces are often subdivided by relocatable partitions. These components can be stored away from the areas in use and brought into place by trolley and lifted by suction pad or scissors cradle on to the track location. Security is achieved by the use of telescopic bolts acting on the top and bottom. The advantage with these methods relies upon skilled management and may be less trouble than using sliding or sliding folding gear where damage can occur with rough usage.

2.2 Basic forms

The basic forms of demountable partitions are:

- stud/sheet systems

- frame/sheet systems
- frame/panel systems
- panel/panel systems
- sliding/folding systems
- sliding relocatable partitions
- screens and office furniture modules

It is useful at this stage to outline the different principles; for examples see sections 2.5–11 inclusive.

2.2.1 Stud/sheet systems (Fig. 2.1(a))

This is primarily a system that depends upon components, cut and fixed together on site. The cladding material is usually plasterboard but cement fibre boards are also used. The industrialized form relies on thin gauge steel studs put together with fast drive screws; the stiffness depends upon the facing boards secured to either faced by screw fixings to the metal frame. The geometry of the layout also contributes. Heights can range from 2.4 to 6.0 m and can be adapted to form lift shaft enclosures up to 18.0 m high. Plasterboard layers can be in duplicate on either face. The hollow core of the studwork can also be filled with mineral wool. Fire ratings of 1, $1\frac{1}{2}$ and 2 hours are customary, but heavier construction with double stud lines can give 3 and 4 hour fire ratings and sound insulation up to 48 dB. Metal stud and plasterboard partitions were developed in the USA 70 years ago. Other versions comprise composite construction that relies upon multiple layers of plasterboard bonded together by plaster dabs, or with a factory-made composite of two plasterboard sheets glued to a honeycomb paper core. Both patterns are limited to domestic heights of around 2.4 m and are likely to be damaged when repositioned. In all forms, services are installed on site.

2.2.2 Frame/sheet systems (Fig. 2.1(b))

This is the most common off-site fabricating system, where the vertical framing is exposed in the form of overlapping trim (using aluminium, plastic or steel). It depends upon steel studs with plasterboard, cement fibreboard, or fire-resisting chipboard panels, fixed to hidden subframing under workshop conditions, the modules being 1.2 or 1.5 m dependent upon the panel material. The systems are easily demountable and can be readily arranged with fanlights/full glazing/sound deadening (up to 25–30 dB reduction)/fire resistance (1 or $1\frac{1}{2}$ hour rating). Heights of 3.0 m–11.0 m can be accommodated. The hollow framing permits pre-wiring and fixing of electrical accessories.

2.2.3 Frame/panel system (Fig. 2.1(c))

The panels can be largely self-supporting so that the members are smaller than frame/sheet studs. The basic principle involves vertical location posts into which a variety of panels, door units or glazing units can be fixed. The frame and panel elements come to site already assembled and are coupled together to form the partition or screen. The post frame is used to accept wiring and accessories, often related to surface trunking at dado or skirting level. Fire rating and sound reduction are similar to frame/sheet systems. A sheet steel system was developed for the Chrysler Building, New York, in 1930 and is still in use today.

2.2.4 Panel/panel systems (Fig. 2.1(d))

The most interesting designs follow the principle of self-sufficient panels that are simply butted together. The support system depends upon top and bottom fixings, often telescopic bolts that relate to floor channels and to grid support at ceiling level integrated with the suspended ceiling system. The distribution of services can be via the panels with bottom and or top connections or through vertical trunking inserted at panel-to-panel junctions. Sophisticated systems available have raised floor components and suspended ceilings on 1.2 or 1.5 m modules as illustrated in Fig. 2.1(d).

2.2.5 Sliding/folding partitions (Fig. 2.1(e))

The most interesting traditional examples have been developed in Japan where paper-faced sliding screens are devised as room dividers with leaves mounted side by side. Such refinements are occasionally imported for Japanese exhibitions, the detailing relying on perfect craftsmanship in hardwood joinery using bottom guides.

European versions rely on metal tracks, usually with overhead support to avoid friction or difficulties with track cleaning. Soundproofing results in heavy construction which requires steelwork or reinforced concrete to carry the tracks above the door openings. Hinging the door leaves with sliding supports at edges or centre-fold position produces a 'concertina' action which can fold away into a 'parking' bay at the side or sides of the opening. The doors can be flush or panelled and be constructed to FD 30 standard and give 25–30 dB sound reduction. The scale and weight of such construction have led manufacturers to produce light versions of folding doors that rely on plastic drapes connected to a lattice frame upon plywood concertina screens with 200 mm leaves. The lighter forms are generally used in domestic construction.

2.2.6 Sliding relocatable partitions (Fig. 2.1(f))

The basis for relocation are overhead tracks for positioning and support. The door leaves can run on motorized tracks

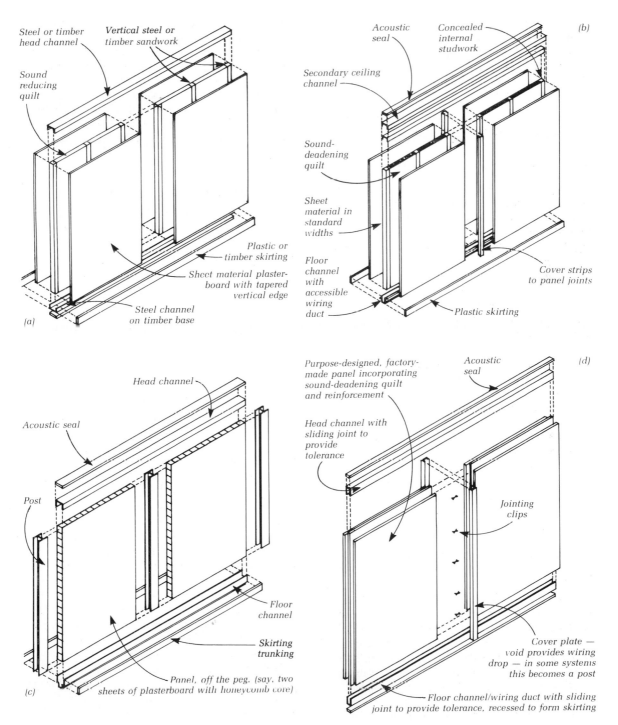

Figure 2.1 Basic forms of demountable partition: (a) stud/sheet system; (b) frame/sheet system; (c) frame/panel system; (d) panel/panel system; (e) sliding folding system (photo: London Wall Design Ltd); (f) sliding relocatable system (photo; The Miro Consultancy); (g) screens and office modules, note combination of framing, furniture and cable servicing (from data for work station designed by Trickett Associates and marketed by Ahrend). *Note*: all board-faced partitions present problems for fixing heavy loads (radiators, sanitary fixings) — see section 2.3.3.

(e)

(f)

Figure 2.1 *continued*

to storage zones or are transferred by fork-lift trucks using suction pad loading. Pneumatic pressure can also be used to inflate seals at the top and bottom of door leaves to improve sound deadening at door edges and give greater rigidity to the construction. Size and weight are not a problem due to mechanical handling, while sound resistance up to 55 dB reduction can be achieved and fire grading equal to 1 hour. Such partitions are commonly used in conference centres and hotel conversions where space precludes storage within the rooms served.

2.2.7 Screens and office modules (Fig. 2.1(g))

The seminal text was Robert Propst's handbook published in 1968 by Herman Miller and titled *Action Office*. This publication revolutionized the concepts for working layouts in health care, laboratories, offices and industry. The open plan notion was tempered by 2030 mm high enclosures with worktops or T-shaped desks set at right angles for face-to-face working conditions. Screens are reinforced by cupboards and the configuration can become cruciform, L or U shaped. The ranges of office systems available today owe a great deal to the pioneering designs by Hille and Knoll International of 20 years ago. Another significant development is the commercial approach to subdivision whereby the building user rents partitions or screens just like office furniture, the owner's responsibility being the raised floor, suspended ceiling and service ducts.

2.3 Performance requirements

BS 5234 : 1992 : Parts 1 and 2 *Code of practice for internal non-loadbearing partitioning* lists properties that may be required and recommends good practice to follow during design, manufacture, erection and maintenance. Reference should also be made to the PSA publication, *Method of Building* 08.101 : 1983.

Partitions — technical guidance The increasing desire for flexible use of spaces within buildings results in a highly competitive market for manufacturers of demountable partitions in which user satisfaction and legislative requirements play a large role. The designer needs precise information about the appropriate use of partition systems from the supplier and data regarding their products where tested to British Standard recommendations. Some systems have British Board of Agrément certificates, while many manufacturers also participate in the BSI Quality Assurance Scheme recommended in BS 5750.

2.3.1 Appearance

Appearance can be a major factor in the selection of a system, and the visual expression provided by frames or joints, as well as the surface finishes of the partitioning itself, are the main areas of consideration. It is worth looking at existing installations to observe how finishes have worn and how easily relocation has occurred.

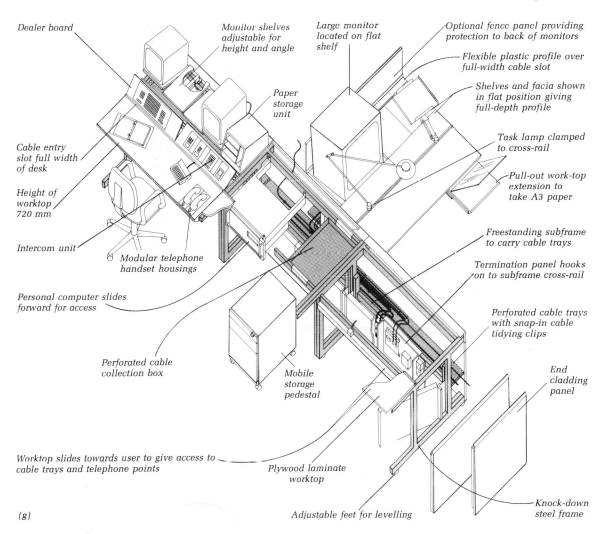

Dealer board

Monitor shelves
adjustable for
height and angle

Large monitor
located on flat
shelf

Optional fence panel providing
protection to back of monitors

Flexible plastic profile over
full-width cable slot

Paper
storage
unit

Shelves and facia shown
in flat position giving
full-depth profile

Task lamp clamped
to cross-rail

Cable entry
slot full width
of desk

Pull-out work-top
extension to
take A3 paper

Height of
worktop
720 mm

Freestanding subframe
to carry cable trays

Intercom unit

Termination panel hooks
on to subframe cross-rail

Modular telephone
handset housings

Personal computer slides
forward for access

Perforated cable trays
with snap-in cable
tidying clips

Perforated cable
collection box

Mobile
storage
pedestal

End
cladding
panel

Worktop slides towards user to give access to
cable trays and telephone points

Plywood laminate
worktop

Knock-down
steel frame

Adjustable feet for levelling

(g)

Figure 2.1 *continued*

Frames are generally of extruded aluminium with anodized or polyester paint finish in a wide colour range, or alternatively painted galvanized steel. Moulded PVC can be used for preformed skirtings, cover strips, trunking and panel edges. The appearance of 'make-up' pieces and junction details should be carefully considered.

Partition wall finishes are usually of two types:

1. *Self finish* (factory applied on preformed panel) which is permanent and requires no maintenance in respect of redecoration, these often include vinyl, timber veneer, plastic laminates, hardboard, steel sheet, fabric (like hessian), glass and acrylic sheet.

2. *Base finish* such as plasterboard, cement fibreboard or chipboard, which will be ready to receive a painted finish. Many systems rely on 'freshening up' with replacement wallpapers, vinyl or hessian facings to the panels. Edge trims that can be loosened and refitted after decoration have a considerable advantage over fixed fittings.

Part of the selection process must be consideration of the preservation of surface characteristics and their resistance to damage by impact or abrasion. Permanent disfiguration can be caused by scorching or chemical action, and finishes can be made unsightly by stains and graffiti. Care must be taken to select finishes appropriate to their relative ease or difficulty in cleaning, redecorating, repair or replacing.

2.3.2 Demountability

The option of demountability is a most significant factor in respect of initial cost, and the effect it will have on the fulfilment of other performance requirements, including appearance. Also, taking down a demountable partition and re-erecting in a new position may involve adaptation to the lighting and heating — although, when the building has been specifically designed with flexible arrangements of accommodation in mind, the services will also allow adjustment without significant alteration of the building fabric. For these reasons demountable partitions are usually employed in conjunction with suspended ceilings and, for maximum flexibility, with raised floors. When considered together, they can provide a grid-based planning flexibility employing compatible construction methods.

The need for cost-effectiveness and efficiency requires analysis of the frequency of change in partition position to be established, because this will determine to a great extent the complexity of the installation and the type of personnel required to carry out the alterations. For example, day-to-day adaptation will require the ease of a sliding/folding partition which can be carried out by the building users themselves; adaptation every few months will allow the use of a more sophisticated partitioning system capable of being manoeuvred by maintenance staff, whereas adaptation after a period of a few years will require the employment of an outside contractor. Properties of sound reduction are significantly reduced in direct proportion to ease of demountability because it will be harder to provide acoustic sealing around edges. A demountable partition system which provides a high degree of sound control will be expensive, both in first cost and cost of reassembly, since the jointing technique will be complex. However, a system which gives demountability only at the junction points, such as doors, abutments and intersections, will be less expensive than a partition which, by reason of more sophisticated jointing techniques, can be demounted at each panel on the planning module. Figure 2.2(a) and (b) show examples of the two alternative methods of obtaining flexibility in demounting:

1. *Detail 2.2(a) 'H section'* The partition panel fits in the vertical 'H' section. The sequence of erection is to fix the wall channel or abutment post and then to proceed, panel—post—panel offered from the side. It is common practice for end panels at junctions with the structure to be odd sizes and be fitted last.
2. *Detail 2.2(b) 'breakdown member'* In this case a 'breakdown' member, which is in three sections so that

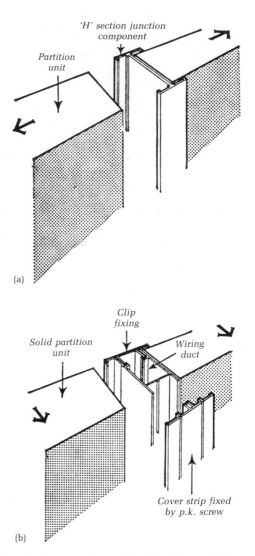

Figure 2.2 Two methods of obtaining demountability.

the panels can be fitted from the front or rear. Demountability can occur at each module.

2.3.3 Partition strength

Although demountable partitions are non-loadbearing they must be capable of lateral stability, and of withstanding temporary sideways point loads as well as soft body impact of prescribed magnitude without deflection. Factors to be considered include the head and base fixings, joints between panels and the strength of the partition construction. Head details are generally most critical since they can be fixed directly to the framework of suspended ceilings. In addition to a rigid head fixing, a framed form of partition, or the insertion of frames in a panel-to-panel form, may be necessary where partition panels are required to carry shelving or other fixtures, including furniture and wash basins.

Manufacturers' recommendations should be taken even when deciding on methods of providing fixings for small items, such as coat hooks or fire extinguishers. British Gypsum provide an excellent source of information in their *White Book* which schedules fixings and fastenings.[1] Suitable methods must support without loosening or damaging the partition and include: back-clamping to the inside of the panel skin (toggles and cavity fixings); threading into fibres or core materials (screws); expansion into the face or core material (nails or screws into plugs, bolts in anchors and expansion bolts); pressure on the face or core material (nails); mechanical clamping through the partition (bolts with washers or backplates on opposite side); and adhesion (glues)

BS 6262 : 1982 *Code of practice for glazing for building* covers the design and performance criteria for vertical glazing within buildings. It is particularly concerned with safety and includes the recommendations of BS 6206 : 1981 *Impact performance requirements for flat safety glass and safety plastics for use in buildings*. See also Chapter 6 on doors. The following sections explore the general problems of fire precautions in subdividing a building or where partition systems may play a crucial role.

2.3.4 Fire precautions in partition construction

When a fire occurs, the danger arises both from within the building and in the risk of the spread of fire from one building to another. The various fire regulations state requirements to limit this spread, the risk of which is related to the use of the building. Factors to be considered include the resistance provided by the constructional elements, the resistance of the surface finishes to the spread of flame, the size of the building, the degree of isolation between the various parts of the building, the provision of fire-fighting facilities and the incorporation of appropriate means of escape for the occupants. Therefore, since demountable partitions are normally associated with building types having a high occupancy, they may have an important role in protecting the people and contents within a building as well as in preventing the spread of fire. Demountable partitions are not specifically mentioned in the Building Regulations 1991. This particular form of partition system is installed within the structural envelope of a building and will not normally be used to provide the fire resistance to the standard of compartment walls (see sections 6.9 and 6.10

for definitions and in particular terms such as stability, integrity and insulation in case of fire). Their 'temporary' nature often makes them unsuitable for use in protecting escape routes. Nevertheless, they must not contribute towards the fire or allow it to spread.

2.3.5 Fire resistance

Most systems will give a notional period of fire resistance of between 30 minutes and 1 hour. In certain situations, demountable partitions may be used temporarily to divide a large fire compartment of a building into smaller areas while certain functions are carried out, for example, a series of separately controlled group meetings within a large hall. In these circumstances, particular care in selection is required in order to make sure that the junctions between panels, and between the partition and adjoining room surfaces, provide the integrity required.

Other factors concerning the ability or otherwise of a demountable partition to contribute towards the fire relates to material combustibility. The ramifications of the 1992 revisions to the Building Regulations in terms of fire resistance are fully explained in detail in section 2.4.

2.3.6 Sound control

A demountable partition by its nature is vulnerable to the passage of sound, and to expect a lightweight site-assembled structure to provide a very high standard of sound control is, to some extent, contradictory.

The matter of sound control should be considered at the design stage of a building so that noise-producing rooms and rooms requiring quiet can be planned remote from each other. In order to solve the problem of protecting the occupants of a room against unwanted noise, both air- and structure- (impact and flanking)-borne sound must be considered as well as external noise (see Mitchell's Building Series (MBS): *Introduction to Building* Chapter 7 Sound Control, and MBS: *Environment and Services* Chapter 6).

Airborne sound is created by fluctuation or waves in air pressure which are perceived by the ear as sound. These waves strike the surface of a partition and cause it to vibrate. In turn this causes the air on the opposite side of the partition also to vibrate and thus the sound is transmitted. The sound-resistant properties afforded by the partition are a measure of the degree to which these vibrations can be reduced or absorbed.

The mass law means that the heavier the material used for a component the greater is its resistance to sound transmission, and implies that demountable partitions are less effective in providing sound resistance than the other forms of partition. However, using lightweight cavity construction, sound reduction values better than those predicted by the mass law can be obtained. For example, a partition constructed of metal studs incorporating a flexible mounting for a plasterboard facing on both sides, results in up to a 10 dB better reduction. The air in the cavity can be considered as a spring between the facings which has inherent damping properties. The effects of this isolation between the facing and the stud can be further enhanced by the introduction of a sound absorbent material into the cavity and are also important behind electric sockets.

Nevertheless, good sound performance in demountable partitions mostly relies on a number of very practical considerations relating to their construction. This must attempt to provide the maximum achievable sound reduction value throughout, including through doors and any infill panels. The design must not permit air leaks caused by joints, gaps, cracks or holes through which sound waves can travel, and in this respect initial workmanship and subsequent handling are of major importance. Potential points of sound leakage can occur around partition edge seals, as well as poorly fitting doors. In some systems, sophisticated constructions, such as resilient air seals at abutments to walls and ceilings, are incorporated to assist in sound reduction as well as to take up possible dimensional inaccuracies in the structure. One irritating source of noise, that of door slamming, can easily be prevented by the fitting of automatic door closers, checks and rubber seals.

Sound waves will also travel over or under the demountable partitions when they are used in conjunction with suspended ceilings and raised floors, especially those incorporating ventilators, heating ducts or continuous lighting troughs. In fact when sound reduction is considered critical and demountable partitions must be used, measures need to be taken that ensure any voids forming part of the construction of the floor, ceiling and flanking walls contain absorbent baffles in positions likely to coincide with a future location of the partition. Alternatively, purpose-made panels must be placed in the cavities every time the partition is moved. This is more difficult for the wall construction than for the suspended ceilings and raised floors.

All sounds differ in pitch and intensity, and BS 2750 *Methods of measurement of sound insulation in buildings and building elements* Parts 1 to 8 : 1980, has been established for comparative acoustic testing. It provides the arithmetical average decibel reduction over the range of 100–3150 Hz (cycles per second) and all partition systems should be tested to this standard. It is possible to produce enhanced performance data by testing over a more limited frequency range, or by testing single panels in isolation rather than a panel and its framing components. In such cases, an apparent gain in sound reduction of 4–5 dB can be incorrectly claimed. Site conditions are very much more difficult than the controlled conditions of the laboratory and, although measured figures are useful for comparison, a poorer acoustic performance on the site is almost inevitable.

In designing for sound resistance, it is important to set targets which are correct in terms of the uses of the spaces separated by partitions. The following list (taken from the British Gypsum *White Book*[1]) indicates the effect of decibel reduction levels on human speech which are useful when selecting a partition system:

20 dB	Normal conversation can be easily overheard.
25 dB	Loud conversation can be heard clearly.
30 dB	Loud speech can be distinguished under normal conditions.
35 dB	Loud speech can be heard but not distinguished.
40 dB	Loud speech can be heard faintly but not distinguished.
>40 dB	Loud speech or shouting can be heard with great difficulty.

Where the background noise level is low, it is generally accepted that an insulation of 40 dB will give a tolerable amount of privacy between two adjacent offices. However, actual performance prediction is dependent on layout, the arrangement of adjacent elements and the presence of absorptive materials. Wherever possible advice should be given by the partition manufacturer.

The changes brought in by the latest revision to the Building Regulations (1 June 1992) provide for mandatory sound control between flats or where residential accommodation is provided next to other building categories and where a separating partition occurs. The likely 'demountable' solution is the formation of sheet and stud construction and where sound reduction will now have to be achieved to the standards required for new construction (usually solid walls of 190 or 215 mm thickness). Alterations where existing half-brick masonry occurs between properties imply additional layers of absorbent materials and drylining to one or both faces of such partitions to obtain similar standards. Part E of Schedule 1 to the Building Regulations stipulates weight of construction, materials and thicknesses.

2.3.7 Integration of services

In partition systems where flexibility is required, service integration is usually restricted to electrical power and communication wiring. Service channels are usually incorporated in suspended ceilings and raised floors, and connection can be made into demountable partitions (other than folding/sliding) by means of special vertical hollow frame units located between panels, or to hollow horizontal dado sections and skirtings. Socket outlets can be fixed in both these vertical and horizontal subcomponents.

Split panel systems are also available which allow one side to be removed independently of the other. These give easy and complete access to wiring for both initial installation and for future maintenance and alteration, and provide greater flexibility for routeing cables and positioning outlets. Metal trunking should be earthed and emerging cables should always be secured and protected by flexible bellows at movable junctions.

2.3.8 Durability and maintenance

Whichever systems of demountable partitioning are employed in a building, it is necessary to take great care during the moving processes in order not to destroy edges or scratch surfaces. Where fairly frequent relocation is necessary a system should be chosen which enables all parts, skirtings, cover strips, moulding, etc. to be reused. Alternatively, it will be necessary to ensure these parts are readily available before relocation is proposed. Most sliding/folding systems can be manoeuvred without damage provided proper care is taken and the tracks and channels are kept in good working order.

Building maintenance manuals should include very precise information about cleaning, including the removal of graffiti, as well as a list of suppliers of replacement components (see MBS: *Introduction to Building* section 15.8 Maintenance team).

2.4 General points on the Building Regulations and fire resistance

The Regulations consider buildings under various purpose groups and which are broadly as follows:

- Residential (dwellings): groups 1a, b and c.
- Residential (institutional): groups 2a and b.
- Office: group 3.
- Shops and commercial: group 4.
- Assembly and recreation: group 5.
- Industrial: group 6.
- Storage and other non-residential: groups 7a and b.

See Schedule AD B, Appendix A, Table A2, in the Building Regulations for amplification. The classification affects compartment zones and their construction and the requirements for fire resistance.

The detailing of partitions is affected where these form compartment walls and where partitions enclose protected escape routes, and in particular the selection of lining materials (both vertical surfaces and ceiling over). Floor finishes are generally exempt since this area is not involved in a fire until it is well established. A critical factor in selecting lining materials for walls and ceilings at risk is the resistance to surface spread of flame as well as non-combustibility. The controlling document is Part B2 of Schedule 1 to the 1991 Building Regulations which requires products to meet certain levels of performance under fire test conditions as laid down in BS 476 *Fire tests on building materials and structures*, also Part 7 : 1971 *Surface spread of flames tests for materials* or Part 7 : 1987 *Methods of classification of the surface spread of flame of products*.

The tests are conducted with a strip of material, one end of which rests against the furnace, the rate of flame then being measured for assessment into categories. Class 1 represents very low flame spread while class 4 represents rapid flame spread and is not acceptable under Part B2, Schedule 1 to the Building Regulations. The other tests look at calorific value of burning materials which result in a higher rate of heat release or earlier ignition. Such materials are hazardous and will reduce time to flash-over in a serious fire. Tests for the rate of heat release are described in BS 476 : Part 6 : 1981 and 1989 *Method of test for fire propagation for products*.

The material is tested for specific times in a furnace and given two indices related to performance. Sub-index (i_1) is derived from the first 3 minutes of test while the overall test performance is signified by the index of performance (I)

Materials passing these more stringent tests are given the classification of class 0, this higher category of performance meaning that linings to class 0 are recommended for high-risk areas such as circulation spaces and escape routes, etc.

Class 0 materials have to comply with one of the following requirements:

1. To be composed of materials of limited combustibility throughout (see below for full definition).
2. To be class 1 material (for resistance to spread of flame) but having indices for tests within furnace as follows:
   (a) index I (not more than 12);
   (b) sub-index i_1 (not more than 6).

2.4.1 Reference term: non-combustible

These are materials which have the appropriate level of performance under fire test and can be defined as any one of the following:

- Tested to BS 476 : Part II and where the material does not flame nor produce a rise in temperature on the furnace thermocouples.
- Be totally inorganic materials (concrete, fired clay, masonry, concrete blocks, etc.).
- Be products classified as non-combustible under tests to BS 476 : Part 4 : 1970.

The use is defined in the Building Regulations, Schedule AD B, Appendix A, Table A6.

2.4.2 Reference term: limited combustibility

This is a more familiar category in the construction of linings to partitions and ceilings and is defined as follows:

- Any material with a density of 300 kg/m^3 or more, which when tested to BS 476 : Part II does not flame and the rise in temperature on the furnace thermocouple is not more than 20°C.
- Any material with a non-combustible core at least 8 mm thick with combustible facings not more than 0.5 mm thick on one or both sides (e.g. plasterboard).

The use is defined in the Building Regulations, Schedule AD B, Appendix A, Table A7, which refers to the use of materials of limited combustibility.

2.4.3 Materials for surface linings

Guidance is given in Table 2.1(a) which is Appendix A, Table A8, taken from Schedule AD B of the Building Regulations. It is useful as a guide to typical performance ratings, but it is important to obtain test results for proprietary materials from makers or trade associations. The overall assembly of components is a significant factor in assessing total performance, and increasingly partition elements are subject to comprehensive fire tests rather than accepting individual figures for separate materials. Significant differences in fire rating can occur if coatings, edge trim, fixings or sealants vary in the final assembly on site.

2.4.4 Thermoplastic materials

These materials relate to ceiling and partition linings in areas that are not for circulation and means of escape and occur more commonly in ceiling construction. See section 3.3.4 for details.

2.4.5 Classification of linings for spread of flame resistance

The recommended classification for spread of flame resistance is given in Table 2.1(b), which is Table 10 (from Schedule AD B2, section 6) of the Building Regulations. The special category for small rooms (class 3) relates to totally enclosed areas and other small spaces or passages (other than protected escape routes).

Certain vertical surfaces are excluded, such as:

- Doors, door frames and glazing in doors.
- Window frames and other frames containing glazing.
- Narrow members, such as architraves, cover moulds and skirtings.
- Fireplace surrounds, mantle shelves and fitted furniture.

Table 2.1(a)   Table A8 (ADB, Appendix A)

Rating		Material or product
Class 0	1.	Any non-combustible material or material of limited combustibility. (Refer to the Building Regulations: Composite Products and Test Requirements, Table A.)
	2.	Brickwork, blockwork, concrete and ceramic tiles
	3.	Plasterboard (painted or not, or with a PVC facing not more than 0.5 mm thick) with or without an air gap or fibrous or cellular insulating material behind
	4.	Woodwool cement slabs
	5.	Mineral fibre tiles or sheets with cement or resin binding
Class 3	6.	Timber or plywood with a density more than 400 kg/m^3, painted or unpainted
	7.	Wood particle board or hardboard, either treated or painted
	8.	Standard glass reinforced polyesters

Notes:
1. Materials and products listed under class 0 also meet class 1.
2. Timber products listed under class 3 can be brought up to class 1 with appropriate proprietary treatments.
3. The following materials and products may achieve the ratings listed below. However, as the properties of different products with the same generic description vary, the ratings of these materials/products should be substantiated by test evidence.

Class 0 aluminium-faced fibre insulating board, flame-retardant decorative laminates on a calcium silicate board, thick polycarbonate sheet, phenolic sheet and UPVC.

Class 1 phenolic or melamine laminates on a calcium silicate substrate and flame-retardant decorative laminates on a combustible substrate.

Table 2.1(b)   Table 10 (ADB2, section 6). Classification of linings

Location	Class
Small rooms of area not more than 4 m^2 in a residential building and 30 m^2 in a non-residential building	3
Other rooms Circulation spaces within dwellings	1
Other circulation spaces, including the common areas of flats and maisonettes	0

The Schedule AD B2, section 6, permits a few variations provided that no lining material of lower category than class 3 is employed. The permitted relaxations are as follows:

1. Wall linings in rooms may be class 2 or 3 if their area does not exceed the following:

   (a)   residential buildings: (A) max. room area 20 m^2;

(B) lining area 50 percent of A or the area employed;

(b) non-residential: (A) 60 m^2; (B) as residential.

2. Rooflights, plastic diffusers and plastic films in the ceiling are considered in section 3.3.4.

3. Internal glazing between rooms must accord with Table 2.1(a) and (b).

High surface spread of flame characteristics are particularly dangerous when occuring within concealed situations, such as in the cavities of partitions, and wherever possible hollow constructions should be avoided in potential fire-risk situations. Alternatively, the internal voids must be limited by the provision of *fire stops*. Manufacturers must always provide enough performance data for an accurate assessment to be made and to allow facts to be presented to the fire control officer.

The concluding paragraphs look at details of varying systems in greater detail and contain some illustrations from completed installations.

2.5 Stud/sheet systems

The systems are extensively used where demountability is the lesser requirement. The advantages rest with the wide range of performance criteria that can be met regarding fire resistance and sound reduction. Doors, clear-storey glazing and windows can be incorporated and it is relatively easy to adapt and alter completed partitions for such features. Heavy-duty constructions can be put together with double studs to fulfil stringent requirements for partitions enclosing compartment zones as Fig. 2.3.

A typical application is employed in fitting out stores for Marks & Spencer where stud/sheet systems with metal framing and plaster/cement fibreboard is used throughout the installation for store partitions (suitably reinforced to carry shelving systems) and for enclosing staff rooms, lift shafts, kitchens and lavatories (suitably finished with washable wallpapers and tiling, etc.).

British Gypsum have industrialized the stud/sheet system, the most useful reference being the *White Book*[1] available from the suppliers which details a variety of applications.

2.6 Frame and sheet systems

A typical example is the Unilock system (Fig. 2.4) and which comprises floor plates and head channels into which are slotted metal studs, with interior studs at mid-panel centres. The facings are 12.5 mm plasterboard in single layer with cores of mineral fibre. A top-hat section in PVC is made to cover the board joints at main stud locations with clip-on mouldings to provide a skirting finish, external corners and at junctions to door frames. A deflection head

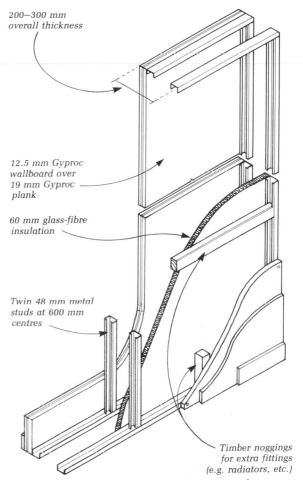

200–300 mm overall thickness

12.5 mm Gyproc wallboard over 19 mm Gyproc plank

60 mm glass-fibre insulation

Twin 48 mm metal studs at 600 mm centres

Timber noggings for extra fittings (e.g. radiators, etc.)

Figure 2.3 Stud/sheet system, heavy-duty construction. Double metal stud separating wall. (By kind permission of British Gypsum Ltd.)

channel allows for sagging lines of a structural floor or ceiling system, plus or minus 15 mm. Details developed for the PSA are fully detailed in Method of Building document MOB 08-137 (February 1989). Fire resistance is from nil to 1 hour, coverings have class 0 resistance to spread of flame and sound-reduction values from 34 to 48 dB, thicknesses are 75, 85, and 100 mm and heights 2.4, 2.7 and 3.0 m.

2.7 Frame and panel systems

Figure 2.5 shows an example of a simple frame and panel system. The frame is of aluminium alloy, anodized to BS 1615 : 1972, and the panels are designed to fit any module using standard sheets. The infill panels are of sandwich construction with a core of polystyrene, or mineral fibre or similar material according to the required

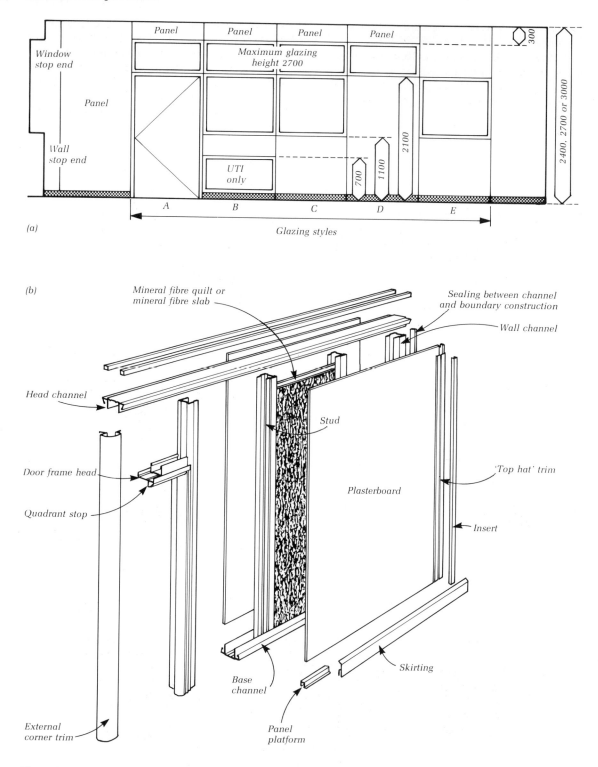

Maximum glazing
height 2700

Window
stop end

Panel

Wall
stop end

UTI
only

Panel   Panel   Panel   Panel

2400, 2700 or 3000

300

2100

1100

700

A   B   C   D   E

Glazing styles

(a)

(b)

Mineral fibre quilt or
mineral fibre slab

Sealing between channel
and boundary construction

Wall channel

Head channel

Stud

'Top hat' trim

Door frame head

Plasterboard

Quadrant stop

Insert

External
corner trim

Base
channel

Panel
platform

Skirting

Figure 2.4 Frame and sheet system by Unilock Partition Ltd: (a) key elevation; (b) assembly details.

performance specification. This core material is faced both sides with hardboard and finished with decorative PVC sheet, decorative laminated plastics or hardwood veneers. More complex panel construction may incorporate a single or double skin of pressed steel or aluminium sheet. This type of double skin panel may also be filled with glass fibre or mineral wool.

The total nominal thickness of the panel illustrated is 50 mm. As with most partition systems, the doors are pre-hung, in this case on nylon washered aluminium butt hinges, and have a rubber strip air seal around the perimeter of the frame. It should be noted that no fire resistance was required in the illustrated construction. Note the use of PVC glazing beads which fit neatly into the 'I' section frame and

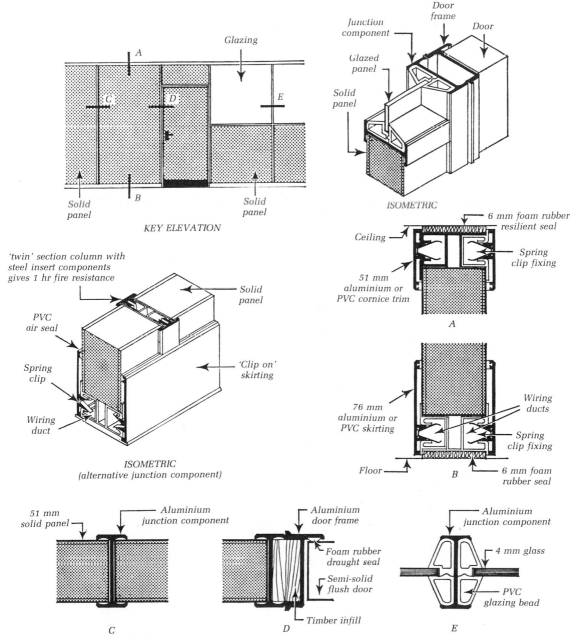

Figure 2.5 Frame and panel system.

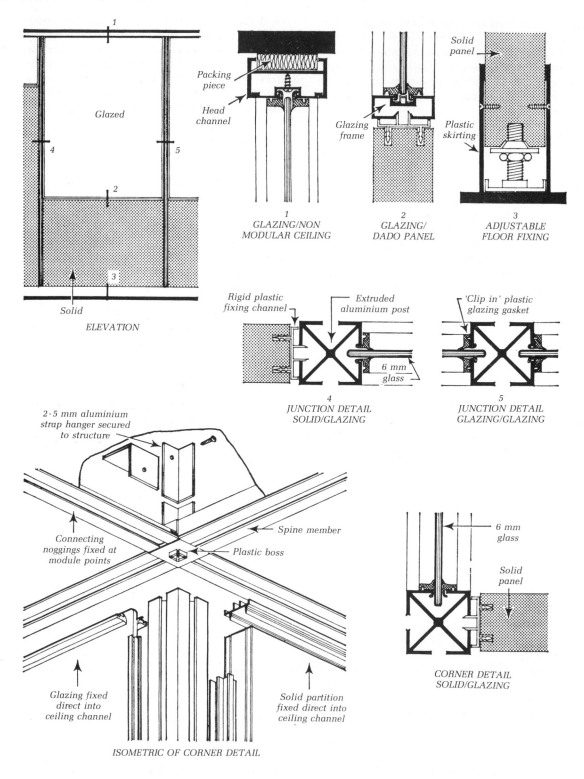

Figure 2.6 Modular ceiling and partition system.

the clip-on cornice and skirting trim. The junction at floor and ceiling is most important, particularly in respect of partitions which must have high sound reduction. In the example illustrated, foam rubber sealing strip is used. Electric and communication wiring can be run in the vertical hollow junction frames as well as along the skirting sections.

2.8 Modular ceiling and partition system

A more sophisticated system is shown in Fig. 2.6. This is a *modular ceiling and partition system* with integrated storage and accessories. The ceiling grid has a dimensional relationship with the partitioning, thereby providing complete flexibility within the discipline of a modular layout. The framework to the ceiling is of extruded aluminium alloy channels suspended from the structure by rod or strap hangers attached to the main span members of the ceiling. Notched and turned aluminium connecting members are fixed to the spine framework at modular points. The junction of the spine and intermediate members is concealed by a plastics boss.

The grid is designed to receive any proprietary acoustic board or tile ceiling. Ceiling and partition modules are standard at 1200 mm. The partition system has a maximum height of 3050 mm, with a 51 mm overall thickness. Panel cores are of mineral fibre, chipboard or expanded polystyrene faced with 3 mm hardboard finished in PVC, laminate or hardwood veneer. The glazing uses extruded plastics beads with glass up to 6 mm thick. Storey height (or normal height) doors are supplied pre-hung and complete with furniture.

One feature of this system is the design of various special brackets which screw or clip into the cruciform framing to provide anchorage for such items as shelves, coat hooks, pinboards and chalkboards. Storage cabinets can also be clipped and hung from the framing, provided that the rigidity of the supports to the ceiling is checked.

2.9 Panel-to-panel system

Figure 2.7 shows a panel-to-panel partition system which is based on a 600 mm modular horizontal planning grid. The preformed monoblock panels are of sandwich construction having two skins of 0.125 mm gauge galvanized steel finished externally with an epoxy polyester powder coat, and have an infill between skins of mineral wool. The overall thickness is 60 mm and the panels can be supplied to suit requirements up to a maximum of 4.5 m high.

The panels are located into cold-formed steel channels fixed to head, floor and end wall locations. The head channel provides a single compartment service wiring duct and the floor channel has a dual compartment for separating electrical and telecommunication cables, access being provided by means of a snap-on 72 mm high PVC skirting.

The vertical junctions between panels are achieved by a grooved edge profile and loose interlocking tongue technique as illustrated. Wall abutments channels allow a ± 10 mm lateral adjustment. Intermediate cold-formed steel posts are used vertically to link glazed panels. Each post has a snap-on pilaster, giving access to a wiring duct. Both internal and external angles can be accommodated using a corner post, and three- and four-way panel junctions are possible.

Where required, panels can be supplied with full height coupled glazing without midrails, and clear or tinted glass is used, 8 mm thick, set in soft neoprene gaskets. Doors in metal frames to match the solid panel construction are available, as well as timber doors in metal frames. They can be either to the same height as the partition panels or 2048 mm high with a matching over-panel. Two panel widths (1200 mm) are necessary where doors occur, and door units are supplied with an adjacent glazed panel to provide continuity with the modular planning grid of the partition system. Both steel and timber doors are fitted with a concealed brush-type draught excluder.

2.10 Cubicle partitions

A special form of panel-to-panel system made for cloakrooms and which encompasses duct fronts and panelled walls which support WCs, wash basins and vanitory units is shown in Fig. 2.8. The materials utilized are described in sections 2.10.1—4 below.

2.10.1 Precast terrazzo

Prefinished slabs of 50 mm made with crushed marble and cement, joined together with dowels, toggle bolts and anchor plates, doors formed in metal or timber and hung on pivot hinges. A typical installation is shown in Fig. 2.8(a). Mesh reinforcement to slabs means that 2.0 m lengths are feasible in storey-height units.

2.10.2 Tile-faced units

Marble or ceramic tile-faced precast concrete slabs, similarly joined together with dowels and anchor plates as terrazzo.

2.10.3 Moulded chipboard panels with laminate facings

Chipboard panels, 25 or 30 mm, with moulded forms for corners and for making vanitory units. Sizes are limited by press moulds used by manufacturers; one-piece elements

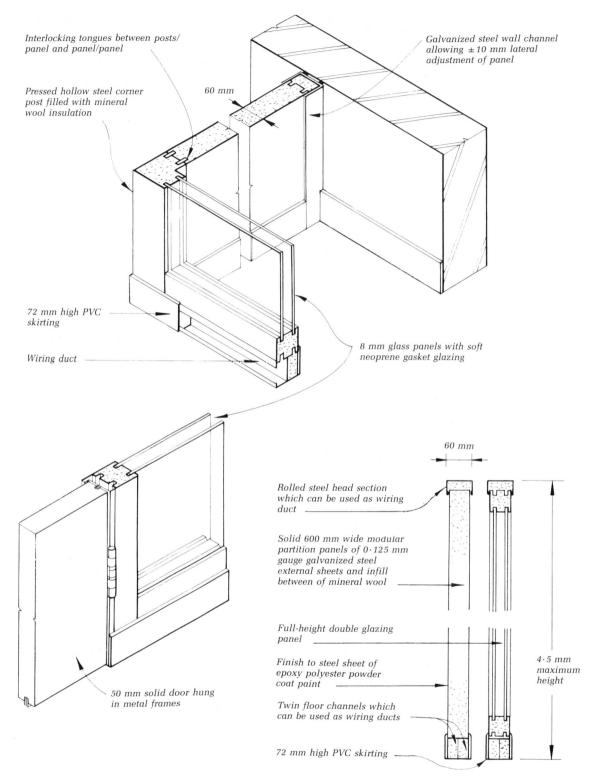

Interlocking tongues between posts/
panel and panel/panel

Pressed hollow steel corner
post filled with mineral
wool insulation

60 mm

Galvanized steel wall channel
allowing ± 10 mm lateral
adjustment of panel

72 mm high PVC
skirting

Wiring duct

8 mm glass panels with soft
neoprene gasket glazing

50 mm solid door hung
in metal frames

60 mm

Rolled steel head section
which can be used as wiring
duct

Solid 600 mm wide modular
partition panels of 0·125 mm
gauge galvanized steel
external sheets and infill
between of mineral wool

Full-height double glazing
panel

Finish to steel sheet of
epoxy polyester powder
coat paint

Twin floor channels which
can be used as wiring ducts

72 mm high PVC skirting

4·5 mm
maximum
height

Figure 2.7 Panel-to-panel system.

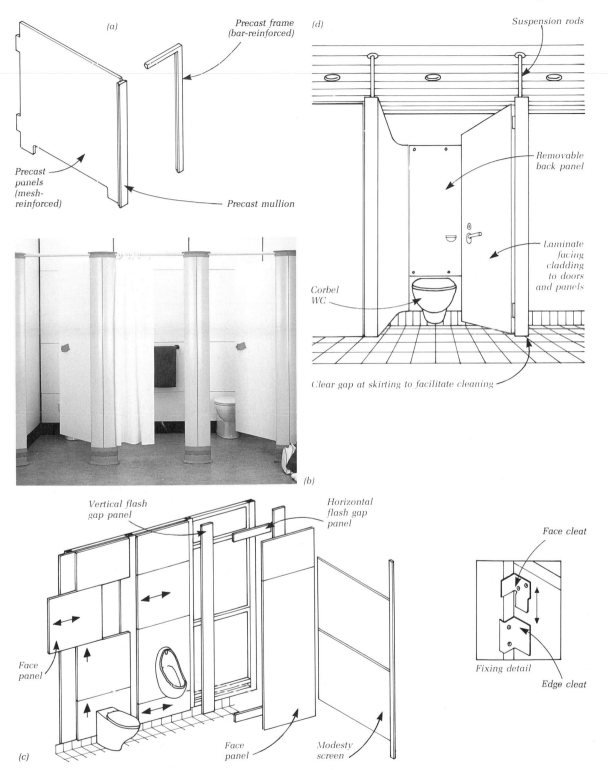

Figure 2.8 Cubicle partitions: (a) precast terrazzo cubicles; (b) and (c) panelled duct and cubicle; (d) cantilever cubicles.

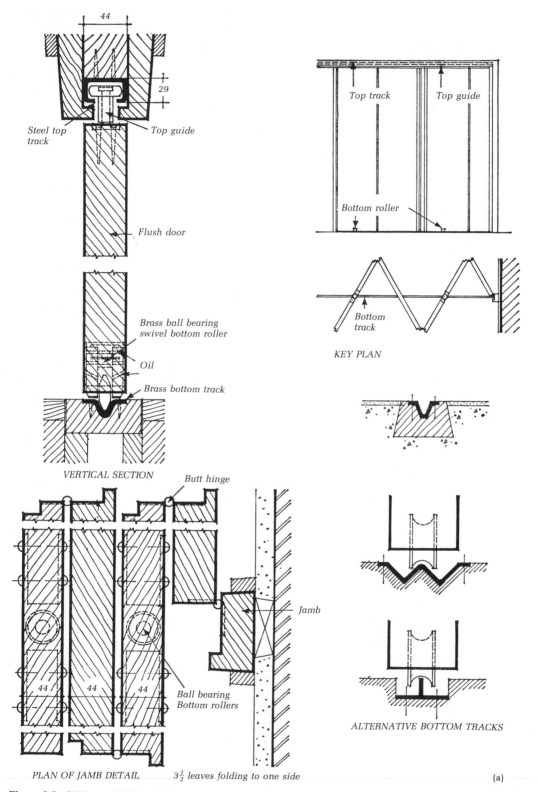

Steel top track

Top guide

Flush door

Brass ball bearing swivel bottom roller

Oil

Brass bottom track

VERTICAL SECTION

Top track

Top guide

Bottom roller

Bottom track

KEY PLAN

ALTERNATIVE BOTTOM TRACKS

Butt hinge

Jamb

Ball bearing Bottom rollers

44   44   44

PLAN OF JAMB DETAIL $3\frac{1}{2}$ leaves folding to one side

(a)

Figure 2.9 Sliding and sliding folding partitions: (a) folding partition; (b) louvred folding partition; (c) collapsible partition.

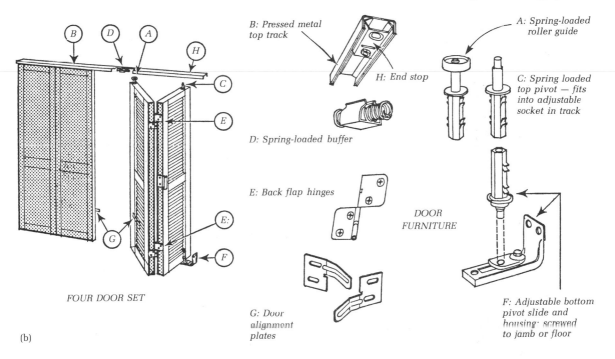

FOUR DOOR SET

(b)

B: Pressed metal top track

H: End stop

D: Spring-loaded buffer

E: Back flap hinges

G: Door alignment plates

A: Spring-loaded roller guide

C: Spring loaded top pivot — fits into adjustable socket in track

DOOR FURNITURE

F: Adjustable bottom pivot slide and housing: screwed to jamb or floor

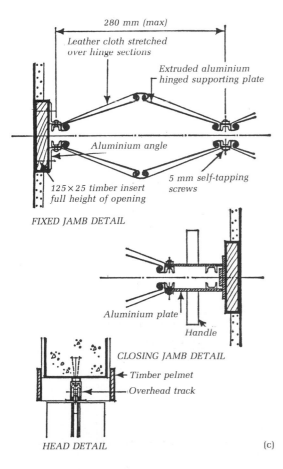

280 mm (max)

Leather cloth stretched over hinge sections

Extruded aluminium hinged supporting plate

Aluminium angle

5 mm self-tapping screws

125×25 timber insert full height of opening

FIXED JAMB DETAIL

Aluminium plate

Handle

CLOSING JAMB DETAIL

Timber pelmet

Overhead track

HEAD DETAIL

(c)

are made up to 2.0 m length and to storey height.

Panelled partitions to rear of WCs and vanitory units are often framed up with steel which is clad with laminate-faced chipboard panels secured to secret bolts to background framing as Fig. 2.8(b) and (c).

Demountable provision is for access to plumbing and water services in ducts behind panelled walls, the release of bolts can be key operated with screwed cover plates as used for mirror fixings. Vandal-resistance finishes include stainless sheet bonded to chipboard.

Ease of cleaning for WC cubicles implies wall corbel WCs and partition divisions having clearance at floor level with tubular leg supports. However, cantilever frames enable moulded chipboard to be carried off rear walls aided by suspension rods from ceiling level, giving total clearance at skirting level (Fig. 2.8(d)).

2.10.4 Moulded plastic

It is also possible to construct cubicles and cloakroom partitions in moulded glass-reinforced fibre with integral epoxy finishes. The elements can have hollow cores filled with mineral fibre and include tubular framing for extra rigidity.

2.11 Folding and sliding systems

2.11.1 Folding

Figure 2.9(a) shows a centre-hung sliding folding partition

used in the construction of a room divider. A half-leaf is necessary against the frame on one side. The particular example illustrates the use of bottom track and rollers with a top guide. The requirements for a bottom track are, however, contradictory since the track giving the least break in the floor surface is least likely to restrain the door properly.

2.11.2 Louvred folding

This type of partition, shown in Fig. 2.9(b), is also used as a room divider or decorative screen. Many manufacturers produce this type as standard at much less cost than purpose-made versions. Decorative timbers such as North American clear pine or Parana pine are used. The rails are dowelled and the louvre slats notched into the styles. This type is also suitable for built-in storage units and wardrobes where maximum access and ventilation is required.

The use of top and bottom pivots leaves the threshold completely clear, and a spring-loaded roller guide is fixed to the top leading corner of each pair of doors to ensure smooth running in the track. Special back-flap hinges are used and the door alignment plates guide the doors together on the closing stile of each pair of doors. If the four panels shown were hung to fold one way, then a different type of track would be required.

2.11.3 Collapsible

This type of partition has almost become standard construction for use as a space divider in houses, and in small community buildings. The partition illustrated in Fig. 2.9(c) is made up of an aluminium alloy collapsible frame over which is stretched leather-cloth, PVC or similar material. These partitions are top hung and do not require a floor fixing or channel, and an important point to note is the minimum amount of space taken up when the doors are folded back.

Note

1   British Gypsum, *White Book*, 1991 Edition, available from: British Gypsum Ltd, East Leake, Loughborough, Leics. LE12 6HX.

3 Suspended ceilings

3.1 Introduction

A suspended ceiling can be formally defined as 'a ceiling hung at a distance from the floor above or from the roof and not bearing on the walls'.[1] These ceilings serve a number of aesthetic functions involving the reduction of room height and provision of decorative planes, as well as practical functions, including the concealment of services, protection of structure against fire, thermal insulation, illumination, heating and ventilation and contribution to sound control.

The trade association which represents installers and specialists is the Suspended Ceiling Association[2] and from that source up-to-date information can be obtained upon installation and manufacture. Many specialists will assemble suspended ceilings to the specifier's requirements by putting together components from different manufacturers; the specifier should therefore be familiar with standard tiles or strip elements and the suspension systems made. BS 8290 : Parts 1, 2 and 3 should be studied regarding standards for setting out and quality control with installations. The methods of construction may vary widely, but can be grouped under the following systems: details are explained in section 3.5 onwards.

- *Jointless systems* These can be 'wet' or 'dry' methods of construction, and provide a joint-free ceiling soffit. They can incorporate removable or hinged panels for access to services located in the space above. A dry membrane ceiling is also marketed from translucent vinyl sheet stretched over aluminium framing.
- *Frame and panel systems* These consist of tiles or planks laid into a framework suspended from the structural floor or roof above. Access is easily obtainable by removing the tiles or planks, although hinged panels may be used at critical positions.
- *Linear strip systems* Similar to the frame and panel systems except that there are many fewer joints owing to the use of long lengths of planks, which may or may not interlock.
- *Louvre and open strip systems* These are formed by open panel or lattice units in a suspension framework. They are used to provide a grille-like decoration to form a visual 'cut-off' from the soffit and services above.
- *Integrated service systems* These consist of a ceiling which forms a service unit for the room below and which can incorporate heating, lighting, ventilation and sound-absorption units. These operate as elements within a tiled panel concept or as metal trays to the whole ceiling area.

3.2 Trade practice

All except 'wet' jointless ceilings are proprietary products. Manufacturers generally work to a 300 or 600 mm module, unless constructional detailing or visual preference makes a 100 mm module (or a 50 mm submodule) more acceptable, such as in the linear strip system. For modular ceilings, the setting out should be carefully considered. Edge cutting with tapered tiles which become slivers look unsightly and are often difficult to secure. Special perimeter infill tiles or strips are usually available so that irregularities in structure can be taken into account, as well as possible movement due to expansion and contraction within the ceiling itself. Trims of painted, coated or capped steel can be used to camouflage ceiling perimeters and junctions. Timber cover moulds or fillets can be applied as a last resort to hide dimensional irregularities at perimeter junctions. Another approach is to dedicate zones of the area for tiled or strip form while the balance is treated in solid manner with plasterboard or fibrous plaster.

The technique for setting out ceiling planes has been advanced by the use of laser beam technology. This is arranged by using a laser tripod near the centre of the building bays which enables a revolving light beam to give a continuous red line for marking out wall battens and for trimming lengths of suspenders. Another use is for the snagging of conduits or ducts which are out of line. The accuracy achieved is ± 1 mm in 30 m. The new BS 8290 reflects these higher standards and contrasts with the previous Code of Practice 290 where lack of alignment equalled 2.5 mm in 12 inches!

3.3 Performance requirements

It is useful to look at an overview of performance from a major user like the PSA. Their publication MOB 09.201 *Technical Guidance, Suspended Ceilings* (dated December 1978, updated 1986) is still the best guide to widely accepted systems and to comparative performance. The PSA also publish useful product data on specific manufacturers, and with references to conformity with BS 476 : Parts 20 to 23 : 1987 (test procedures and fire resistance, etc.), also BS 476 : Part 7 : 1971 *Surface spread of flame* and BS 476 : Part 8 : 1972 *Criteria for fire resistance*, as well as BS 8290 : Parts 1, 2 and 3 on quality control.

Other British Standard references are as follows:

- Materials and their testing:
  BS 1230 : Part 2 *Gypsum plasterboard.*
  BS 1470–1477 *Wrought aluminium and aluminium alloys.*
  BS 2989 *Hot dipped galvanized plain sheet and coil.*
  BS 2972 : 1975 *Methods of test for inorganic thermal insulation* (includes thermal performance, fire hazard, vapour performance).
  BS 2994 *Cold rolled steel sections.*
- Dimensional systems and tolerances (both documents are essential reading for suspended ceiling design):
  BS DD 22 *Co-ordination of dimensions, tolerances and fits for buildings.*
  BS PD 6440 *Accuracy in building.*

It is important to consider the ceiling in relation to any partitioning system that may be used in the building. In some cases the use of a raised floor system can also have an influence on the type of suspended ceiling, particularly when the building incorporates a large number of mechanical services (see Fig. 4.1).

3.3.1 Appearance

Suspended ceilings are often used for the sole purpose of improving the appearance of the underside of the floor or roof above. They may be required to conceal service pipes or the visual results of certain construction techniques, such as composite floors formed with steel decking to the soffit, which are not considered compatible with the functions to be carried out in the space below.

A lowered ceiling can be used to improve the function and quality of interior spaces by altering the height proportion in relation to the width and length. In this respect the positioning and size of access and lighting panels and/or ventilation grilles play an important role in the overall pattern effect created by the ceiling in relation to other features in the room. The ceiling itself could be translucent and lit from behind, or be solid with a smooth surface or a range of textured surfaces. Auditoria and other spaces can be given directional emphasis by a sloping or profiled soffit through the use of suspended ceilings. Fibrous plaster is used for decorative ceiling work where moulded profiles are required, and sections up to 6 m square are possible using a suspension system similar to that employed for wet plaster construction.

The setting out of modular ceilings and the junction detail between ceiling and wall are particularly important as are the quality and accuracy of workmanship used during their installation. See also comments above regarding the use of perimeter trims in metal, plaster or timber.

3.3.2 Weight

The self-weight of the suspended construction plays a critical role in choosing suspension members. The engineering of the structural floors may well dictate limitations for superimposed loads, fitting-out operations with raised floors, suspended ceilings and partition systems. The jointless ceilings using traditional techniques of plaster and render are the heaviest form of suspended ceiling, weighing between 20 and 50 kg/m^2 (Table 3.1 and see Fig. 3.3). By contrast, PVC film is a very light form of construction, weighing between 3 and 5 kg/m^2. However, the construction must always be designed to be strong enough to support the weight of lifting and ventilation units incorporated in the ceiling, and this loading should always be calculated for each individual scheme. It is worth while recalling a collapse at the Piccadilly Plaza Hotel, Manchester, where a substantial area of the banqueting room ceiling fell down just before the guests arrived for the opening occasion.

3.3.3 Accessibility

Accessibility is essential where the space between the suspended ceiling and roof or floor space over is used as a horizontal service void for heating pipes, water and waste pipes and where all types of wiring and cable work can be easily and freely run. Thus, the question of easy access to the space for maintenance purposes is of primary importance. It is important that the design of the suspended ceiling not only takes into account the provision of an adequate number and size of access points, but also ensures that the suspension points do not conflict with those required

Table 3.1 Typical examples of opaque and luminous suspended ceiling materials and their weights (BSCP 290 : 1973, Tables 1 and 2A)

Material	Thickness (mm)	Approx weight (kg/m²)
Opaque		
Expanded metal and plaster	40	20−50
Wood or other organic fibre insulation board	12.7	3.7
Mineral fibre and wool insulation board	12.7−15.9	4.9−6.1
Perforated steel tray with insulation	30 o/a	7.3−14.6
Perforated aluminium tray with insulation	30 o/a	3.7
Gypsum plasterboard	12.7	10.25
Gypsum insulating plasterboard	9.56	7.5
Gypsum fibrous plaster tiles	15.9−31.8	14.7−17.0
Glass fibre tiles and boards	19.0	3.7
Luminous		
Louvred polystyrene		4.9
rigid PVC		4.9
metal		0.5−5
Closed diffuser		
PVC sheet		2.3
acrylic/PVC		2.3
polystyrene		3.7
PVC film		4.9
glass fibre		4.9

for the services above. A system of design work using overlap techniques by drafting, most likely to be computer-aided, will ensure that pipework, ventilation ductwork, electrical trunking, fire barriers, acoustic barriers, loudspeakers, signs and recessed light fittings are fully co-ordinated with the ceiling suspension in the ceiling void. Adequate space allowance must be provided within this void for installation as well as future maintenance of services and, for example, when air-conditioning must be incorporated in 750 mm depth of service zone.

A modular panel, which can be removed over the whole area of the ceiling, becomes the easiest and most complete means of access. The consideration of the direction of the services, and the amount of connection required to the vertical ducts and to partitions services, etc. determine the amount of access required and will help to decide the type of suspended ceiling most suitable. Narrow strips which can be removed over certain areas will give access to parallel service runs, if necessary right up to the partition, while jointless ceilings will be satisfactory where a few predetermined access points are acceptable.

A ceiling system which gives full access may lead to an undisciplined design of the service runs. There is also the risk that the ceiling may be damaged if too many panels are removed. Where very extensive or frequent access is needed, a proprietary hinged access panel should be used, or a washable panel from which fingermarks can be removed.

Access panels may present fire test problems; checks may be needed on airtightness problems. The hinged trap, though expensive, has better performance standards under test conditions.

3.3.4 Fire precautions

The requirements for suspended ceilings in the context of fire precautions are divided into two areas of consideration: fire resistance of construction and surface spread of flame characteristics.

Fire resistance There are some circumstances where a suspended ceiling contributes to the overall fire resistance of the structural floor above. Typical situations occur in the upgrading of old buildings or where change of use implies higher fire standards. In these circumstances the ceiling construction must fulfil the criteria for fire resistance and implies jointless materials (cement and sand rendering and plaster over expanded metal) or multiple layers of plasterboard (aggregating to 30 mm thickness). Both methods can provide 1 hour of fire resistance. Consideration must be taken of the height of the building, whether or not the floor is a *compartment floor* and the total period of fire resistance required.

The inference from these requirements is that the type of suspended ceiling relying on separate tiles within a framework of metal angles or 'T' sections is not allowed as contributing to the fire resistance of the total floor construction in buildings more than 15 m high where the period required is 1 hour, or in the buildings of any type where the period required is more than 1 hour. In the latter circumstances, the ceiling must be of jointless construction.

Where a ceiling does not have to be fire resistant, combustible materials can be used. However, panels of combustible material falling to the floor or on to furniture can increase the spread of fire. Certainly all ceiling fixings must be non-combustible, which implies steel hangers or suspenders as well as steel bearers and runners. Notwithstanding the above requirements, once penetrated by fire the ceiling void can become a horizontal flue and assist the spread of fire and smoke to another area of the building, either on the same floor or to the floor above should gaps occur in the floor construction for ducting and vertical pipework. The ceiling void must, therefore, be adequately compartmented by the use of fire-resisting barriers, and compartment walls should be taken up through the ceiling to the level of the structural soffit. Gaps in floors

(a)

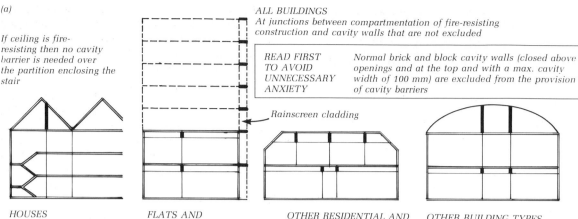

If ceiling is fire-
resisting then no cavity
barrier is needed over
the partition enclosing the
stair

ALL BUILDINGS
At junctions between compartmentation of fire-resisting
construction and cavity walls that are not excluded

READ FIRST TO AVOID UNNECESSARY ANXIETY	Normal brick and block cavity walls (closed above openings and at the top and with a max. cavity width of 100 mm) are excluded from the provision of cavity barriers

Rainscreen cladding

HOUSES

- Above the enclosure to a protected stairway in a house of 3 or more storeys (can be a fire-resisting *ceiling or a cavity barrier in the roofspace*)

FLATS AND MAISONETTES

- In voids above fire-resisting walls to escape routes
- At floor levels and at compartment walls behind rainscreen cladding in buildings with top floor over 20 m (aD has been corrected (15 m should read 20 m)

OTHER RESIDENTIAL AND INSTITUTIONAL

- Above bedroom partitions
- Above corridor enclosures
- To subdivide cavities so that there is a maximum dimension of 20 m between barriers (or 10 m when not class 0 or 1)
- Above corridor subdivision (see other building types)
- Behind rainscreen cladding (see flats and maisonettes)

OTHER BUILDING TYPES

- In voids above enclosure to protected escape routes
- Above (smoke) doors dividing lengths of corridor
- To subdivide cavities (see other residential and institutional)

for ducts and pipes should be carefully sealed; a number of intumescent products are available for this purpose which form a fire barrier under heat conditions to ensure fire resistance at compartment zones. The same process applies to fire barriers within ceiling voids as explained below.

Fire barriers in concealed spaces (including suspended ceilings)   Refer to Appendix E of the *Building Regulations 1991: Approved Document B*. It should be explained that many buildings constructed today contain hidden voids or spaces within floors (either raised floors and or suspended ceilings) walls and roofs. This is particularly true with framed buildings which may contain combustible wall panels, fixtures as well as electrical services, thereby increasing the risk of unseen smoke and flame spreading through concealed spaces. Therefore, despite compartmentation and the use of fire-resistant framing, many buildings have been destroyed as a result of fire spreading through horizontal or vertical ducts which bypass compartment floors and walls.

The basic approach is to interrupt the cavity so that a fire shaft is not formed through compartment zones and to subdivide the voids if they are very large. Appendix E describes the acceptable arrangement for cavity barriers constructed to restrict the movement of smoke or flame within concealed spaces and cavities. These are summarized by illustrations in Fig. 3.1(a) with references to the various categories of building type. A more detailed examination

(b)

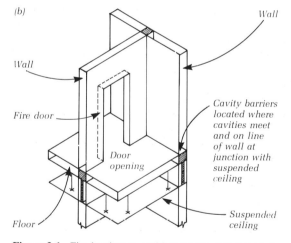

Wall

Wall

Fire door

Door opening

Floor

Cavity barriers located where cavities meet and on line of wall at junction with suspended ceiling

Suspended ceiling

Figure 3.1   Fire barriers to cavities: (a) illustrated guide to arrangement of fire barriers (after 'Easiregs' by Henry Haverstock in *Easibrief* Morgan-Grampian (Construction Press) Ltd, 1993); (b) closing of cavities (junctions at ceilings, floors and hollow partitions).

of the junctions involved is depicted in Fig. 3.1(b) with cellular floors and partition framing adjoining a suspended ceiling. The class of surface exposed in the cavity is stipulated in Schedule AD B3, section 9, and summarized for non-domestic buildings in Table 3.2. Provisos are made which allow the 20 m dimension to be increased to 40 m if the following restrictions are made:

Table 3.2 Non-domestic building cavity classification (based on schedule AD B3)

Location of cavity	Class of surface in cavity (but excluding pipes, etc.)	Max. dimension in any direction
Between roof and ceiling	Any class	20 m
Any other cavity	Class 0 or 1	20 m
Any other cavity	Class 2 or 3	10 m

- Room and cavity space to be compartmented from rest of building.
- Automatic fire detection and alarm system fitted.
- AC duct work and venting ducts, etc. to accord with BS 5588 : Part 9.
- Exposed ceiling within cavity to be class 0 and fixings to be non-combustible.
- Pipe insulation to class 1.
- Electrical wiring carried on metal trays or be in metal conduit.
- Any other materials in cavity to be of limited combustibility.

It may be difficult to construct an effective fire barrier within the ceiling void where the void is congested with services, and it is often necessary to group ducting and pipework to reduce the need for individual cutting and packing at each service pipe or duct. Access panels in fire-rated suspended ceilings must be treated in the same way as fire doors, and for proprietary systems the manufacturers' recommended details should be closely followed. Finally the Building Regulations recognize that service pipes may have to pass through cavity barriers and lay down specific details for such situations.

Surface spread of flame The ceilings can constitute a considerable fire hazard in respect of their surface spread of flame characteristics. Therefore, the *Building Regulations 1991: Approved Document B2* places controls on the type of surface use for a ceiling, depending on its precise location within a particular purpose group of buildings. See section 2.4 for purpose groups and to Schedule AD B, Appendix A, Table A2, in the Building Regulations for amplification. The selection of materials (class 1, 2, 3 or 0) for ceiling surfaces is similar to the selection of linings also described in section 2.4. There are the same references already described to BS 476 *Fire tests on building materials and structures* with subsection references as follows: Part 4 : 1970; Part 6 : 1981 and 1989; Part 7 : 1971 and 1987. Note also that the terms already used in discussing partitions in Chapter 2, apply to ceilings, namely indices *i* and *I*, 'non-combustible' and 'limited combustibility'. See Table 2.1(a) and (b) for typical performance rating of materials (where relevant to ceilings) and to Table 10 (Schedule AD B2, section 6) for classification of linings to ceilings.

Thermoplastic materials The Regulations allow concessions for thermoplastic materials used for suspended ceilings which do not meet the required surface spread of flame characteristic discussed in section 2.4 Table 2.1(a) and (b)). Thermoplastic materials form an integral part of ceiling systems either as lighting diffusers or as rooflights perforating the ceiling zone. Their inclusion depends upon the following criteria:

- Compliance with classification in Table 10 (Schedule AD B2, section 6) (see Table 2.1(b)).
- Thermoplastic material with a TP(a) rigid classification.
- Plastics with a class 3 rating or else type TP(b) numerated above.

There are limitation in layouts for (b) and (c) categories as given in Fig. 3.2 and with more general rules (as in Table 3.3(a)) for all categories of lighting diffusers and rooflights within suspended ceilings.

Further requirements for ceilings Fire protection in suspended ceiling construction can be used to contribute towards the fire resistance of a floor. The criteria for this aspect are given in Table 3.3(b) (namely Table 3 from AB B, Appendix A of the Building Regulations). The differing forms are as follows:

- *TP(a) rigid* Means rigid PVC sheet or solid polycarbonate (not double/multi-skin) at least 3 mm thick or multi-skinned (uPVC or polycarbonate) with a rating of Class 1 (BS 476 : Part 7). The other forms are rigid thermoplastic which complies with tests under BS 2782: Part 5 : 1970 (1974) *Method 508A*. These are commonly used in lighting diffusers and rooflights.
- *TP(a) flexible* These are flexible products, no thicker than 1 mm, which comply with type C of BS 5867: Part 2 : 1980 and to the tests stipulated in BS 5438 : 1976, Test 2 1989. They are usually films used as lighting diffusers or ceiling panels.
- *TP(b)* Usually rigid polycarbonate sheet less than 3 mm thick or the multi-skin form, neither complies with standards for TP(a). These and any other products between 1.5 and 3 mm thick which comply with tests in BS 2782 : Part 5 : 1970 (1974) *Method 508A*. Materials are commonly used for rooflights.

There are four categories of ceiling, types A, B, C and D. Type D is the highest grade and should be constructed of materials of limited combustibility and should not contain

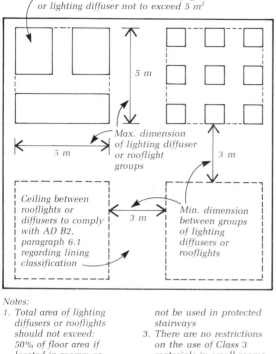

Maximum area of individual rooflight or lighting diffuser not to exceed 5 m²

5 m

Max. dimension of lighting diffuser or rooflight groups

5 m

3 m

Ceiling between rooflights or diffusers to comply with AD B2, paragraph 6.1 regarding lining classification

3 m

Min. dimension between groups of lighting diffusers or rooflights

Notes:
1. Total area of lighting diffusers or rooflights should not exceed: 50% of floor area if located in rooms; or 15% of floor area if located in circulation spaces
2. Plastics rooflights and lighting diffusers should

not be used in protected stairways
3. There are no restrictions on the use of Class 3 materials in small rooms
4. There are no restrictions on the use of TP (a) materials except for note 2 above

Figure 3.2 Limitations on use of class 3 plastic rooflights, TP(b) rooflights and TP(b) lighting diffusers.

easily openable panels. This type is used where the fire resistance of the total floor or ceiling assembly exceeds 60 minutes. Types A, B, C and D have to comply with surface spread of flame requirements of Table 2.1 (a) and (b). It should be remembered that maintenance of services calls for access areas in the suspended ceiling and these are best arranged as proprietary hinged panels with tested fire resistance rather than depending upon unskilled tradesmen dismantling and putting back areas of ceiling tile to gain access to service areas, thus leaving the fire resistance impaired.

3.3.5 Sound control

Suspended ceilings can contribute to the sound resistance between floors and adjoining rooms, and also correct acoustics by absorbing or reflecting sound. Whereas the former relies on mass and airtightness, the latter is a property of the construction or surfaces of the materials employed.

Although the structural floor is the main sound barrier, the ceiling can be used to contribute towards the overall resistance provided. The weight of the supporting structure, depth of cavity between floor and ceiling and the weight of tiles or panels are significant factors. When heavy construction is used for the supporting structure, say reinforced concrete of 200–400 kg/m², the effect of a relatively lightweight suspended ceiling (see section 3.3.1) will be insignificant when considering sound transmission between floors. It follows that, if the supporting structure is lightweight, the suspended ceiling can make an effective

Table 3.3(a)   Table 11 (AD B2, section 6). Limitations applied to thermoplastic rooflights and lighting diffusers in suspended ceilings, and Class 3 plastics rooflights

Minimum classification of lower surface	Use of space below the diffusers or rooflight	Maximum area of each diffuser panel or rooflight*	Max. total area of diffuser panels and rooflights as percentage of floor area of the space in which the ceiling is located	Minimum separation distance between diffuser panels or rooflights*
TP(a)	Any except protected stairway	No limit†	No limit	No limit
Class 3‡ or TB(b)	Rooms	5 m²	50%	3 m²
	Circulation spaces except protected stairways	5 m²	15%	3 m²

* Smaller panels can be grouped together provided that the overall size of the group and the space between one group and any others satisfies the dimensions shown in Fig. 3.2.
† Lighting diffusers of TP(a) flexible rating should be restricted to panels of not more than 5 m² each, see paragraph 6.14 of AD B2.
‡ There are no limitations on class 3 material in small rooms.

Table 3.3(b)   Table A3 (AD B2, Appendix A). Limitations on fire-protecting suspended ceilings

Height of building or separated part (m)	Type of floor	Provision for fire resistance of floor (minutes)	Description of suspended ceiling
Less than 20	Not compartment	60 or less	Type A, B, C or D
	Compartment	Less than 60	
		60	Type B, C or D
20 or more	Any	60 or less	Type C or D
No limit	Any	More than 60	Type D

Notes:
Ceiling type and description:
A Surface of ceiling exposed to the cavity should be class 0 or class 1.
B Surface of ceiling exposed to the cavity should be class 0.
C Surface of ceiling exposed to the cavity should be class 0. Ceiling should not contain easily openable access panels.
D Ceiling should be of a material of limited combustibility and not contain easily openable access panels. Any insulation above the ceiling should be of a material of limited combustibility.

Any access panels provided in fire-protecting suspended ceilings of type C or D should be secured in position by releasing devices or screw fixings, and they should be shown to have been tested in the ceiling assembly in which they are incorporated.

contribution to the overall sound resistance provided it is as heavy as practicable, dense in composition, imperforate, not too stiff and isolated from the structure above by flexible supports.

The resistance to sound provided by the ceiling is critically important when considering transmission between adjoining rooms. Sound will travel through tiles or boards via joints in the suspension system or through pores in the material. For this reason an acoustic tile ceiling can have a sound resistance as low as 12 dB. This figure can be vastly increased by using a jointless ceiling (a 12 mm plasterboard ceiling gives 28 dB, a double-layer system 33 dB), or by laying an absorbent quilt over the tiles.

Suspended ceilings are commonly used to provide the correct acoustic requirements in a room. The machinery used in an open-plan office, such as printers, computers and photocopiers, require the use of acoustic absorbent material which can only be conveniently placed at ceiling level. CP 290 specifies four types of acoustic absorbent:

1. *Resonant panels* to absorb sound near their resonance frequency, normally between 50 and 200 Hz — they include materials such as plywood and hardboard, although lighter-weight panels have a resonance frequency of up to 500 Hz.
2. *Porous surface panels* to absorb frequencies of 500–4000 Hz — they include material such as mineral fibre.
3. *Semi-perforate and perforated composite panels* having a porous surface material which is fissured, textured, perforated, drilled or slotted to increase acoustic efficiency and improve appearance.
4. *Perforated panels* backed with independent acoustic

absorbent materials used over an airspace combine the advantages of porous absorbents with resonant panels.

From the above it will be seen that frame and tile and linear strip systems with 'open' joints and perforations with overlay quilts provide the best acoustic absorbers. Acoustic control can be provided over selected areas of the room to suit particular conditions. Painting mineral fibre tiles and perforated metal tiles when redecoration falls due will impair their acoustic performance.

The acoustics of the space must be considered as a whole, taking into account wall finishes, carpets and furniture. Whenever possible, partitions should be continued into the suspended ceiling void (see section 2.3.6).

Where sound absorption is not a requirement, open grid or strip ceilings can be used as these are acoustically transparent and allow the full volume of a space to be used to increase reverberation time. This could improve the acoustics for speech or avoid the room becoming acoustically 'dead'.

3.3.6 Durability and maintenance

Many of the problems associated with suspended ceilings occur because insufficient consideration was given to their subsequent maintenance. Faulty access panels or the simple removal of modular tiles can be a major source of client complaint unless the design of the ceiling takes into account the realities of service maintenance, some of which may occur in an emergency situation.

Building maintenance manuals should include very precise information about cleaning, as well as a list of suppliers of replacement components (see MBS *Introduction to Building* section 15.8 Maintenance team).

3.4. Jointless systems

3.4.1 Expanded metal and plaster

Continuous jointless surfaces may be required for fire resistance or where ceilings have to be provided with a durable and tough surface, The usual backing is expanded metal with galvanized or stainless steel hangers fixed to the overhead structure. Careful layout and specification for plaster mixes are needed to ensure a fully bonded and crack-free finish.

Spray coat finishes have been developed for mesh where acoustic properties are needed, but it should be noted that such absorbent plaster may lead to a breakdown of the bond between metal and the render base coats in hostile corrosive environments.

There is another form of jointless ceiling using galvanized wire reinforced by vermiculite dabs, formerly called Colterralath, which withstood 25 years of weathering at the Battersea Pleasure Gardens. More recent is Twil-lath of wire mesh and weatherproof paper backing, normally used for renovation work.

The choice of lath is crucial. Expanded metal provides the best guarantee of stability, but the choice of metal and type of plaster is critical if corrosion is to be avoided. Dovetailed or keyed patterns have greater strength, the heavier grade galvanizing giving good on-site protection. The design intention of these ceiling meshes is that cement and sand render coats are forced into the lath so that the rear face is totally encased in cement. Fixings are bitumen-coated galvanized steel or stainless steel.

These forms of lath are structural and will span 1800 mm between supports. The large-scale use of this technique can be seen on London's Underground stations where stainless Halfen channels are fixed clear of the old station walls to support curving arches of Hy-rib for new rendered and tiled surfaces.

Expanded metal and plaster are relatively cheaper than the other ceiling systems, although the slowness of the wet construction is a disadvantage.

The ceiling shown in Fig. 3.3 is a jointless suspended ceiling consisting of plaster applied to metal lathing. Two methods of securing the hangers to support the lathing are shown, and Fig. 3.4 indicates the construction of an access panel. This construction method can be used to provide ceilings to almost any moulded shape according to design requirements, limitations only being provided by practical considerations relating to the bending of the lathing and space for the application of the plaster. Light fittings, ventilation ducts etc. can be easily accommodated provided they are planned in advance of construction. The tough surface can be enhanced by using cement—lime—sand render finishes and when the material is suitable for protected surfaces such as canopy soffits. Galvanized laths should be used in damp environments.

Using 15 mm thickness lightweight vermiculite plaster,

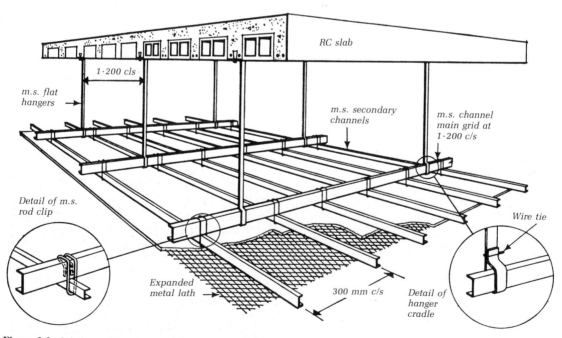

Figure 3.3 Jointless ceiling systems: plaster on expanded metal lath.

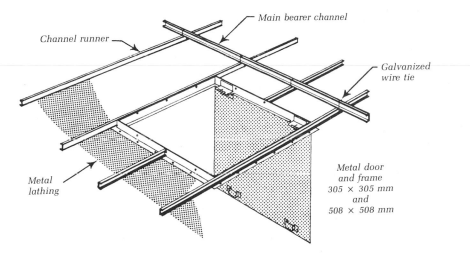

Figure 3.4 Jointless ceiling system: access door.

a jointless ceiling system of this type will weigh 20 kg/m^2, including the suspension. It will have a fire resistance of between 2 hours[3] or more, up to 4 hours when a greater thickness of plaster is used and a class 0 surface spread of flame classification.

3.4.2 Metal firrings and plasterboard

This is a quicker method of construction than expanded metal and plaster, giving fire ratings of up to 2 hours. It is not recommended for areas of high humidity or externally. Figure 3.5 illustrates typical construction which can be modified to two-layer plasterboard and with a backing of rockwool to increase sound deadening. Skim to plaster can be applied, but dry finishing with grouted joints is usual; overlay with lining paper or textured coatings will hide joints.

3.4.3 Fibrous plaster

This is the oldest form of ceiling and has long been used for the repetitive features like the enrichments of Adam ceilings. Catalogues from such specialists as Jacksons (G. Jackson & Sons, founded by George Jackson 1766–1840 and now part of Clark & Fenn Ltd) or Aristocrat illustrate the standard items made from their stocks of master moulds. Ceiling fixers specialize in prefabricating large panels which can be made up to 6 m square and incorporating wire reinforcement, suspenders, lighting and ventilating ducts. The advantage of fibrous plaster rests with its dimensional stability in a dry environment, enabling panels to be jointed *in situ* with minimal cracking. Figure 3.6 illustrates a typical installation.

The more customary application is in tile form, and the best ranges of acoustic and air-handling units are made in this manner. Standard 300 × 300 mm and 600 × 600 mm series exist, but a non-standard format is economic for areas

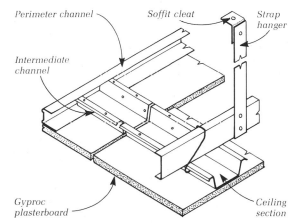

Figure 3.5 Jointless ceiling with plasterboard screwed to metal framework. (By kind permission of British Gypsum Ltd.)

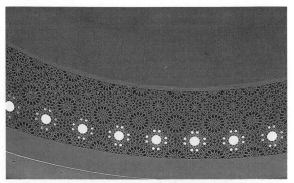

Figure 3.6 Typical installation of fibrous plaster ceiling.

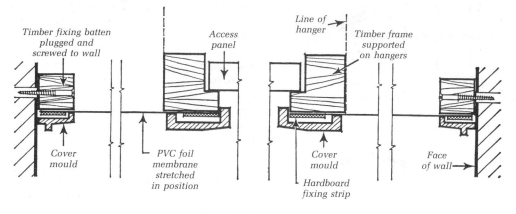

MEMBRANCE CEILING

Figure 3.7 Jointless ceiling system: PVC membrane.

of more than 500 m^2. The tiles can be fixed as lay-in units with a visible grid, but it is more usual to employ secret fixings because tile edges are sufficiently robust to allow removal without damage.

It is easy to achieve decorative features. Any colouring is a matter of painting, in which case the cellular pattern for air-handling grilles needs to be sufficiently 'coarse' to prevent infilling by overpainting, say a minimum of 5 mm openings. Glass-reinforced gypsum is an alternative.

3.4.4 Membrane ceilings

Figure 3.7 shows details of the use of a PVC foil membrane, stretched to form a jointless suspended ceiling. This type of ceiling can be installed in one sheet 0.2 mm thick, forming a panel up to 7.5 × 6.0 m. The method of construction is adaptable to any shape of room, and will accommodate any size of access panel or light fitting. The finish of the PVC foil is semi-matt, and does not require decoration. Where a higher degree of sound absorption is required, a special perforated foil with a loose backing of absorbent quilt is used.

Membrane ceilings are comparatively cheap but give no fire protection (see section 3.3.4 for comments about fire resistance and the use of plastics materials for suspended ceilings).

3.5 Frame and panel systems

3.5.1 General points

This is the most common type of suspended ceiling, comprising a light structural frame of metal angle or tees supporting infill panels or tiles of plasterboard, plywood, fibre building board or mineral wool or, alternatively, metal trays or plastics sections. For details of the sheet sizes and physical properties of these materials, see MBS: *Materials* as well as *Finishes*. Proprietary panels are usually modular co-ordinated to 300, 600, 900, 1200 or 1800 mm square; some suppliers make rectangular panels.

There are various components used for securing the framing for the infill units as given in key plan form for Fig. 3.8:

Hangers Metal straps, rods or angles which hang vertically from the main floor or roof construction to support and 'level-up' the suspension system. Various methods of securing the hangers to the main floor construction are shown in Fig.3.9.

Bearers These are the main supporting sections connected to the hangers and to which the subsidiary horizontal runner supports are fixed. The use of bearers enables the hangers to be at wider centres than the basic ceiling module. There are many ingenious proprietary methods of attaching the runners and hangers.

Runners These are the supporting members which are in contact with the ceiling panels. They are usually of aluminium T- or Z-sections. The runners span in the opposite direction to the bearers.

Noggings These are subsidiary cross members which span in the same direction as the bearers, but in the same plane as the runners in order to complete the framework. Runners and noggings can be concealed or exposed.

Fixing methods Typical methods of securing the ceiling panel into the suspension system are shown in Fig. 3.10(a)−(d) and are described as follows:

(a) *Exposed fixing* Here the ceiling panel or tile drops into the suspended framework formed by extruded

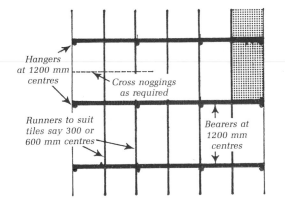

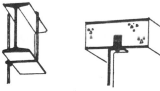

Hangers at 1200 mm centres

Cross noggings as required

Runners to suit tiles say 300 or 600 mm centres

Bearers at 1200 mm centres

Figure 3.8 Frame and panel system: plan layout of suspensions.

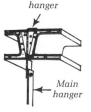

Hanger rod clipped around joist

Strap hanger plugged to concrete

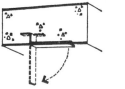

Fishtail strap hanger cast in concrete

Strap hanger screwed to side of wood beam

Stub hanger

Main hanger

Stub hanger fixed between precast units

Strap hanger clipped to BSB

Figure 3.9 Frame and panel systems: hangers.

aluminium T-sections. Unless there is a risk of the panels being lifted by wind pressure (e.g. in an entrance hall), they can be left loose and are thus very easy to remove where the void above is used as a duct. Where they are required to be held down, a wire or spring metal clip is slotted into the web of the T.

(b) *Concealed fixing* This is a concealed type of fixing since the method of support is not visible from below. The figure shows the grooved tiles slotted into the Z-section runner. An alternative form of concealed fixing using a tongued and grooved tile is illustrated in Fig. 3.11. The concealed framework makes access more difficult, particularly at the head of partitions.

(c) *Clip fixing* In this detail a special runner is used which holds a metal tray in position. The tray will be perforated and will probably have an infill of mineral wool or similar inert, non-combustible, sound-absorbent material.

(d) *Screw fixing* This detail shows two alternative forms of securing the ceiling panels by direct screw fixing.

3.5.2 Advantages and disadvantages of frame and panel systems

The erection of a frame and panel ceiling is speedy and clean, and this type of ceiling is usually easily demountable. However, they are often too lightweight to be effective in terms of sound resistance, and the large number of joints makes them less able to resist fire and smoke penetration. Frame and panel ceilings are also more vulnerable to damage during installation and subsequent maintenance than other systems. Nevertheless, they can be easily integrated with overhead services, and parts of the supporting frame may be constructed of larger metal sections which act as air intakes/extracts for air-conditioning plant. The framing can also be used to support light fittings, incorporate lighting track for spotlights, or be the connection point for fire alarms and sprinkler systems. Some manufacturers offer fully integrated service systems, including an air-diffusing framework, and modular lighting units.

The visual effect depends upon the type of fixing grid — exposed semi-concealed or concealed. Exposed grids are more flexible and cause less problems with broken tile edges when changes are made. Concealed fixings give a monolithic appearance to the ceiling, and heavy textured tiles make the joint lines disappear.

There is a standard edge trim to cope with wall or column junctions and to give either a 'flash gap' or simple angle trim below the tile. The present trend with mineral fibre tiles is to use self-coloured material instead of white, and

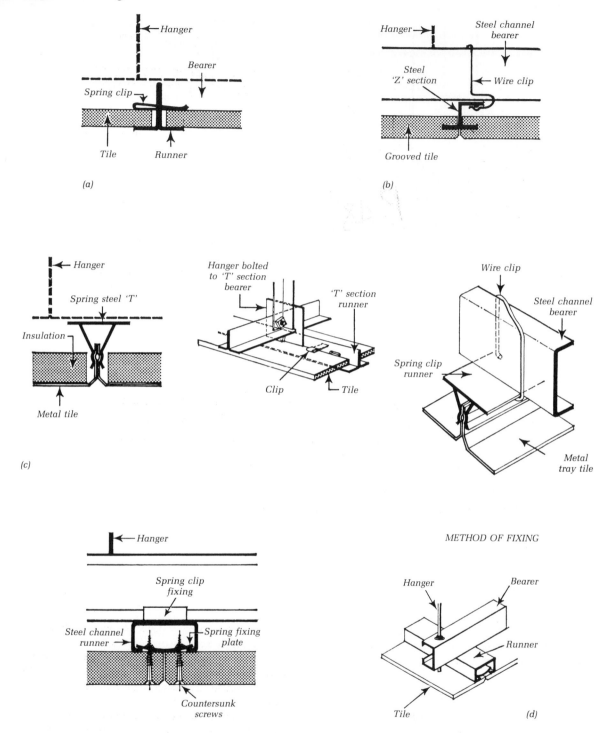

Figure 3.10 Frame and panel systems — typical fixing methods for ceiling systems: (a) exposed; (b) concealed; (c) clip; (d) screw.

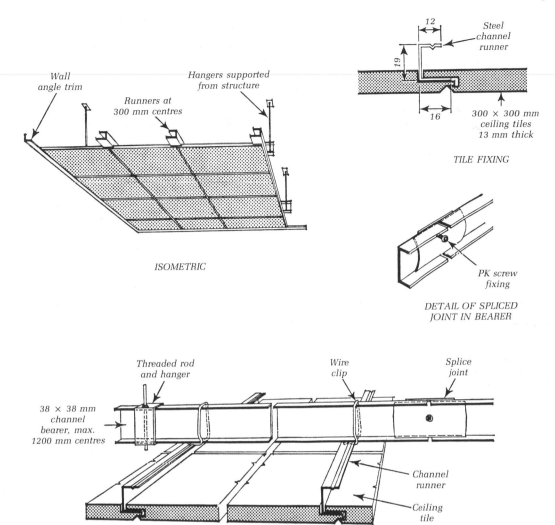

Wall angle trim

Hangers supported from structure

Runners at 300 mm centres

ISOMETRIC

12

19

16

Steel channel runner

300 × 300 mm ceiling tiles 13 mm thick

TILE FIXING

PK screw fixing

DETAIL OF SPLICED JOINT IN BEARER

Threaded rod and hanger

Wire clip

Splice joint

38 × 38 mm channel bearer, max. 1200 mm centres

Channel runner

Ceiling tile

SUSPENSION SYSTEM

Figure 3.11 Frame and panel systems: concealed fixing details for ceiling systems.

one can expect this to develop into a wide colour range.

The industry is using colour and visible grid patterns to overcome the complaint of 'anonymity' or 'tile by the mile'.

3.6 Linear strip systems

The type of ceiling shown in Fig. 3.12 uses material in long lengths in order to minimize the points of suspension and provide a linear visual effect. By using the inherent rigidity of metal and timber in strip form, the ceiling need only be supported in one direction. For example, fixing points can be as much as 6.000 apart for deeply profiled metal

sheets using a suspension system hung at 1.800 centres. However, most linear strip systems have support grid modules of up to 1200 × 1500 mm and the width of exposed metal strips varies upwards from 100 mm. The strips are available in a range of polyester coating colours as well as a mirror finish.

In all other respects, this system fulfils similar performance criteria to those of the frame and panel system. Little fire resistance is provided unless the strips interlock, and access is obtained by removal of several strips, which can be inconvenient because of their length.

Refinements in detail can be made by alternating lighting

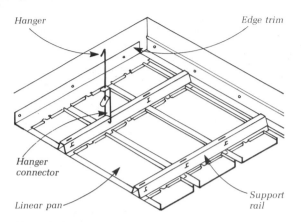

Figure 3.12 Linear strip system with metal strips clipped to notched support rails. (By kind permission of Donn Products (UK) Ltd.)

and heating services in continuous strips between linear panels. Coated steel strips are made that interlock to improve fire resistance and which can be overlaid with mineral wool. Perforated patterns also exist to improve sound resistance, and there are also heated systems which include hot-water pipe runs clipped to the rear face.

3.7 Louvre and open strip systems

The type of ceiling shown in Fig. 3.13 consists of a series of strips or louvre panels made from timber or metal which are supported within as a suspended mesh or trellis design. They provide a visual cut-off when seen from below and help to conceal service pipes and unsightly construction details. They can be lit from above and there is a wide range of fittings available to fit into various cell sizes, these giving diffused effects as well as spot lighting.

Sound absorption is not good, but it can improved by using deep louvres of perforated metal with a quilted core. However, the reverberation time of a space can be increased by allowing the full volume of the room to be used despite its partial division by the suspended ceiling. Although this system provides no fire resistance, the open ceiling can be useful where smoke extraction is needed and where sprinklers can be placed *above* the ceiling plane. The single set of sprinklers are thus effective for the floor space and for the service voids within the ceiling area. This type of semi-open ceiling has most advantage where there are a large number of services to which access is frequently required, but whose appearance is considered unacceptable for full exposure.

3.8 Integrated service systems

Suspended ceilings in which the services form an integral part of the construction (Fig. 3.14) fall into the following categories:

- *Fully illuminated ceilings*, having infill panels of translucent plastics which form the diffusers for the light fittings suspended above them. The diffusers may be plain faced, three-dimensional, corrugated, embossed or 'eggbox' construction. This type of ceiling should not be confused with the use of modular light fittings, which fit into the grid spacing of the panel and frame suspended ceiling system.
- *Ceiling panels incorporating low-temperature heating*, using electrical elements or small-bore cooled or hot water circuits, normally in conjunction with sound-absorbing panels (as Fig. 3.14.(a)).
- *Plenum chamber ceilings* in which the whole of the space between the ceiling and floor or roof soffit is used for

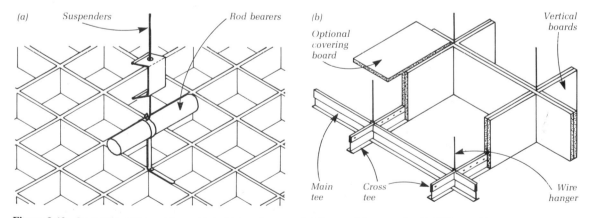

Figure 3.13 Open grid ceilings: (a) moulded chipboard, suspended from rod framework; (b) 1200 × 1200 module with vertical boards as baffle, 1200 × 1200 spaces can contain lighting or ventilation units or be open.

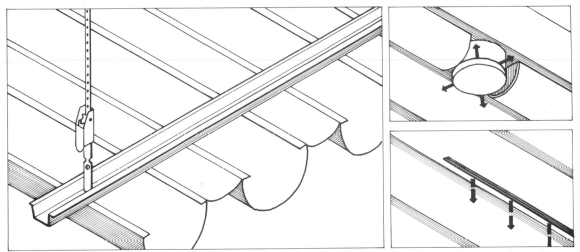

In the ceiling different forms of ventilation units can be built in, either as ventilation diffusers to the same profile as the ceiling units, or semi-recessed circular units

(a)

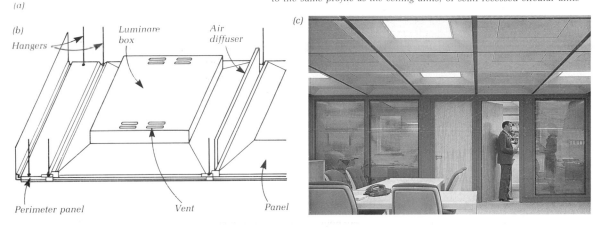

(b)

Hangers

Luminare box

Air diffuser

Perimeter panel   Vent   Panel

(c)

Figure 3.14 Integrated service systems: (a) ceiling panels incorporating low-temperature heating; (b) fully serviced integrated ceiling (by kind permission of *The Architects' Journal*); (c) integrated ceiling and relocatable partition system (by kind permission of Clestra Hauserman Ltd); (d) suspended ceiling with cooling or heating pipes above perforated metal trays (by kind permission of Grayhill Blackheat Ltd (Frenger Ceilings)); (e) Lloyd's Building, London, key section showing integration of services with structure and raised floor elements (architects: Richard Rogers partnership).

the circulation and direction of hot or cold air (see Fig. 4.1).

- *Fully 'serviced' integrated ceilings* which give an element of construction providing cooling, heating and lighting, as well as sound absorption. This arrangement, though expensive, permits the ultimate in flexibility of use for the space below (see Fig. 3.14(b), (c) and (d)).

Integrated service system ceilings have a much larger grid layout than the other ceiling systems, ranging from 900 × 900, 1200 × 1200, 1500 × 1500 to 1800 × 1800.

There are also examples where the structure is integrated to the services with a waffle slab that accommodates lighting and heat-extracting units, the electrical, electronic and air

output units being housed above the floor slab within raised floor components. An outstanding example is Lloyd's Building London designed by Richard Rogers Partnership Ltd (Fig. 3.14(e)). Other buildings where the structure provides the ceiling role without overlays are offices such as Lloyd's, Chatham, and CEBG, Bedminster Down, both designed by Arup Associates.

Notes

1   A useful source for definitions in building is BS 6100 : *Glossary of building and civil engineering terms* Part 1 *General and miscellaneous* subsection 1.3.3 1984 Floors and ceilings. Many

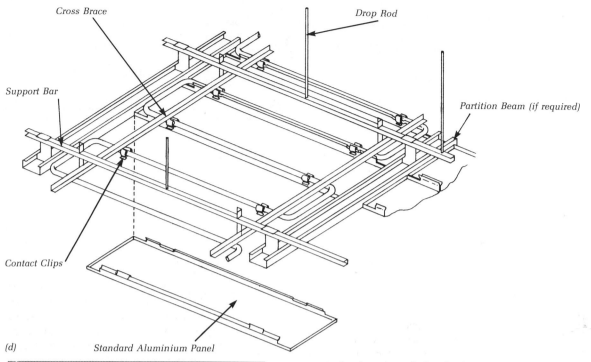

Cross Brace

Drop Rod

Support Bar

Partition Beam (if required)

Contact Clips

(d)   *Standard Aluminium Panel*

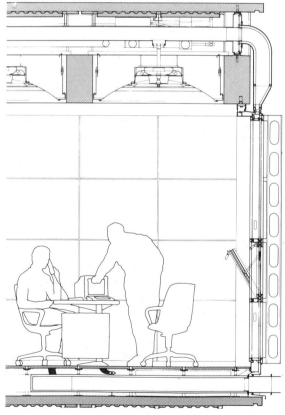

(e)

aspects give the author a feeling that 'newspeak' is springing off the page.

2 Suspended Ceiling Association, 29 High Street, Hemel Hempstead, Herts HP1 JAA. An excellent source of information concerning good practice and advice on installation and manufacture of components.

3 Specification for 2- and 4-hour fire resistance. Typical specification for 2-hour fire rating to a beam and floor soffit. Steel beams encased with Expamet BB 252 attached to 25 × 3 mm flat or 6 mm mild steel rod stirrups at 350 mm centres, fixed by clipping or welding to steelwork or Rawlbolting to con rete slab. To receive 13 mm thickness of vermiculite—gypsum plaster and Expamet angle beads to arrises. Typical specification for 4-hour fire rating to a beam specification as above, but allow for 32 mm thickness of vermiculite—gypsum plaster. Angle bead to be applied after rendering by the plasterer.

Figure 3.14   *continued*

4 Raised floors

4.1 Introduction

Modular floors which are constructed above the level of the structural floors on which they rest are known as *raised floors*. Raised floors were originally developed as a response to the design requirements for rooms accommodating mainframe computers 30 years ago. In those days, cooling systems and festoons of cables could be conveniently located in a floor void and where ready access for their maintenance was required. Since those developments there has been an increasing need for the provision of raised floors as a standard fitting for general office area, as well as other spaces where cable services play a vital role in their function (Fig. 4.1). The developing technology is not without pitfalls — excess heat from office equipment and poorly arranged cooling systems have plagued older installations. There are also factors such as inadequate care of air-conditioning plant with risks to health and comfort. Traditional solutions sought to provide large quantities of fresh air needed from the direction of the ceiling and which results in uncomfortable down draughts. Current ideas include the concept of local cooling through recirculating air-conditioning cabinets with the servicing for these, including pipework, within the raised floor system. Clearly, the substitution of the underfloor space as a recipient area for services instead of the suspended ceiling requires considerable forethought in the way service grids are organized and the way space capacity is incorporated.

It is, perhaps, the most significant change in the fitting of old buildings that has occurred in the past 20 years. The electronic office today depends upon work stations connected to a variety of cable systems, together with electrical supplies and pipework. The loading carried by the raised floor is classified as in Table 4.1.

In North America underfloor ducts are more popular than raised floor systems because of possible savings in building height which may be as much as 600 mm per storey. This can make a substantial saving with a 90-storey block. Casting ducts within the screed on slab is also less expensive than for raised floor components which may have a life cycle as short as 5 years! Because of their wide use in the USA there are many specialist firms experienced in connecting into duct systems to install equipment. In the UK no comparable service exists and work usually revolves around builders' work to make and change connections while the office struggles to continue business.

Duct systems in the UK are fabricated from aluminium or steel and run on a cellular basis (electrical, telephone and data) with or without service outlets. Service outlets provide the user with a convenient access, say, at 1200–1500 mm centres. Leaving out the outlets means drilling through the screed each time a service connection is made. Tolerances must be considered between structure and metal duct systems, say 10 mm to allow for bedding and fixing lugs (Fig. 4.2).

In many buildings today, the precise needs of work stations and their layout are decided late in the fitting out. It is easier to accommodate final adjustments for extra cooling, additional cabling and desk-mounted task lighting via raised floor components than via partition or suspended ceiling systems. Screens between work stations can be treated as furniture, hence the growth of this division of partitioning at the expense of traditional forms.

There are fundamental questions to ask concerning raised floors before decisions can be reached:

- How accessible does the floor have to be?
- What depth of void for services is required?
- What level of fire resistance is needed, if at all?
- Which quality and type of finish is required? Remember that some systems have a tough, prefinished panel in their range that is specific for heavy traffic zones and machine areas.

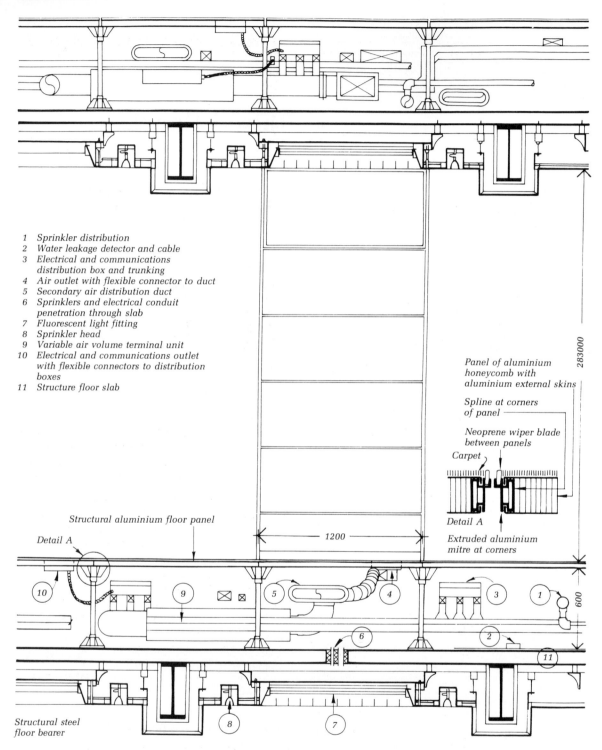

1   Sprinkler distribution
2   Water leakage detector and cable
3   Electrical and communications
    distribution box and trunking
4   Air outlet with flexible connector to duct
5   Secondary air distribution duct
6   Sprinklers and electrical conduit
    penetration through slab
7   Fluorescent light fitting
8   Sprinkler head
9   Variable air volume terminal unit
10 Electrical and communications outlet
    with flexible connectors to distribution
    boxes
11 Structure floor slab

Panel of aluminium
honeycomb with
aluminium external skins

Spline at corners
of panel

Neoprene wiper blade
between panels

Carpet

Detail A

Extruded aluminium
mitre at corners

Structural aluminium floor panel

Detail A

1200

283000

600

Structural steel
floor bearer

Figure 4.1 Section through typical raised floor at Hong Kong Shanghai Bank. (Architects: Sir Norman Foster and Partners.) Note that the raised floor void contains all service runs.

Table 4.1 Raised floor load classifications

	Overall loads	Concentrated point loads
Heavy loading	12 kN/m²	4.5 kN per 25 mm²
Medium loading	8 kN/m²	3.0 kN per 25 mm²

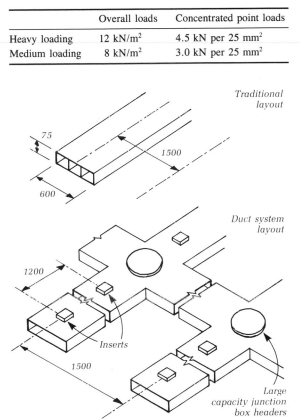

Figure 4.2 Duct system on 1500 mm grid.

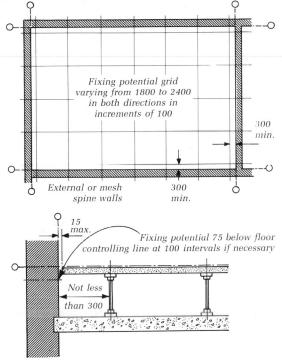

Figure 4.3 Typical layout plan and section.

• How many accessories will be needed from manufacturers, for example ramps, steps and guard railing?
• Is the building design suitable for raised floors? Namely loading capacity, levelness of structural subfloor, clear heights from proposed raised floor to ceiling, also impact upon levels in terms of stairs and lift thresholds and upon sill level (sill level to finished floor under section 20 of Buildings in Inner London). Details of regulations concerning Section 20 buildings can be obtained from the London District Surveyors' Association, PO Box 15, London, SW6 3TU — these relate to buildings of excessive height or volume within the confines of the former GLC area in London.

There are two categories of raised floor construction:

1. *Shallow systems* These comprise panels laid over timber battens, or a metal decking fixed to the subfloor and provide a cavity below of less than 100 mm.
2. *Deep systems* These comprise panels supported by adjustable metal jacks which provide a cavity below of at least 100 mm, and are particularly useful for the accommodation of an extensive amount of services as well as when the structural floor surface is not level.

For shallow raised floor systems, the panels are made from steel, steel-faced particle board, or plywood; and for deep raised floors the same material can be used as well as fibre-reinforced inorganic materials, cast aluminium and glass reinforced cement (GRC). Non-rigid floor panels can give a springy, hollow sensation when being walked across. The panels can be self finished if appropriate, or finished in a range of sheet materials, including carpet and PVC. In a computer room, a material must be chosen which will not store static electricity, and metal-framed raised floors should provide a simple earthing mechanism.

Common modular sizes for panels are: 600 × 600, 750 × 750 and 1200 × 1200 mm. Typical plans are given in Fig. 4.3.

4.2 Performance requirements

As yet there is no British Standard which gives recommendations about raised floor systems. However, the PSA have produced useful guides, the most useful being

Platform Floors (raised access floors) MOB PF2 PS/SPU dated 1992. These are also reports on specific installers covering Robertson floors (March 1990), Hewetson floors (June 1988 ref. MOB 08 714) and System floors (August 1988 ref. MOB 08 716).

4.2.1 Appearance

The two factors affecting the appearance of raised floors are the visibility or non-visibility of the modular grid of the panels used for their construction, and the type of material used for the surface of the floor.

Whereas the modular grid of a suspended ceiling and, perhaps of a demountable partition, will not necessarily influence the desirable positioning and the size of furniture, etc. in a room, a strongly emphasized floor grid could give aesthetic problems relative to the degree of dimensional incompatibility. If this is likely to be considered a problem, then the designer must ensure that the flooring system integrates satisfactorily with the furniture. However, most raised floors can allow surface abutment of floor finishes at panel junctions which will not over-emphasize the modular grid. In this case, care must be taken to ensure compatability between the modular grid of the floor panels (usually 600 mm) with that of the finishes: carpet tiles favour 500 mm modules and office planners work to a 1200 or 1500 mm grid. A successful solution to this problem relies on co-operation between the different suppliers and users. Wall-to-wall floor finishes which lay over all the panels can be used, but for deep raised floor systems, will not provide the intended ease of flexibility for access to the services below.

Window sill and transom levels should be designed to suit the range of potential floor levels provided by the raised floor.

The constructional technique for flooring to non-serviced areas comprises decking units or joists and blockwork sleeper walls to save the cost of pedestals and trays. Far more important are the related floor levels for lobbies, stairs and communal service areas such as lavatories, kitchens and door thresholds.

4.2.2 Loading

The loading capacity of a raised floor is determined by the type of support, and panel material, thickness and area. Loadbearing capacities of between 1.5 and 3.0 kN/m^2 and distributed loads of up to 12 kN/m^2 are typical for most proprietary raised floor systems. Shallow floors are suitable for lightweight use only and can take normal general office loading of 2.5 kN/m^2; heavy equipment, such as safes, large computers and bulk storage, need special consideration. The use of high-grade panels and jacks allows deep raised floors to take greater loads than shallow floors, and many systems can mix these components to facilitate localized loads. The provision for heavy computer loadings are often determined by the requirements demanded by computer manufacturers.

4.2.3 Accessibility

Shallow and deep raised floor systems are used to provide different degrees of accessibility. Whereas shallow systems are generally used where regular access to the concealed services is not required, deep systems are best used where:

* a large proportion of the floor needs to be accessible;
* services need to be routed across deep-plan buildings;
* the floor void serves as a plenum for heating and ventilation.

As the shallow systems are simpler and cheaper than the deep systems, a decision must be made about the depth of void required and whether or not every panel is to be removable. For some servicing requirements, it may be possible to provide only point access at fixed service outlets, for permanently fixed work stations on an open-plan office floor.

4.2.4 Fire precautions

Raised floors are described as *platform floors* in the *Building Regulations 1991: Approved Document B* and are not required to have any specific fire resistance. Nevertheless, for the purposes of safety, and in order to limit the spread of a fire, certain regulations require the whole of any raised floor to be made of non-combustible materials. Accordingly, only non-combustible materials should be used within the floor void and timber battens must be treated with a fire retardant to improve safety. In critical situations, panels must be of a cement-based composite board rather than be wood based. As the void of a deep floor system may represent a fire hazard owing to the presence of a great number of cables, cooling plant, etc., it may need its own fire detector and protection systems (including sprinklers) in high-risk areas. It is also prudent to provide drainage facilities to allow water used for fire fighting to escape.

Where partitions are forming fire compartment walls, they should be carried down through the raised floor if it is not fire resistant. Horizontal voids in the floor may also need to be subdivided by fire/smoke stop barriers (section 3.3.4).

Airtightness is another factor if the underfloor void is to be used as a supply plenum. It is recognized as better practice to duct the supply air to avoid dust pollution, the underfloor space then being used for a return air plenum and subject to filtration before recirculation. Under the latter condition air sealing to the tray system is not important and does not impose high standards of fit between floor trays when laid, or after being shifted or refitted.

4.2.5 Sound control

Raised floor systems are seldom rigid; noise can therefore be transmitted through the construction and permeate adjoining rooms. For this reason, fixed partitions should always be continued through to the structural floor below, and in certain cases sound-absorbent material will be required in the void to reduce the build-up of noise levels. The presence of excessive amounts of moisture can cause floor panels made from timber products to swell and distort which makes them creak when walked across or vibrated by machines. For this reason ground-floor slabs must be adequately water- and vapourproofed, and in doubtful situations only panels unaffected by moisture should be specified.

4.2.6 Durability and maintenance

Like most components, the materials employed and skill of installation as well as subsequent usage all affect the

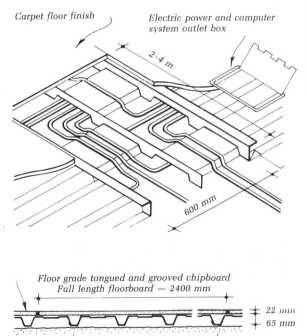

Figure 4.4 Shallow raised floor system in galvanized steel fluted trough decking. (By kind permission of H. H. Robertson.)

durability and maintenance of a raised floor system. Frequency of access is an important factor as panels can become damaged and, in extreme cases, subject to wear. Some manufacturers provide purpose-designed carpet-lifting tools, and also double-suction lifting tools for ease in removing panels for access (see Fig. 4.4). One of the advantages of using modular carpet or PVC tiles is that they can easily be replaced.

4.3 Shallow systems

Shallow raised access floor systems generally have a small cavity, not exceeding 100 mm, which is capable of accepting electrical, communication and data cables. There is little provision for adjustment, and a very smooth structural floor is needed. Removable covers are provided at prescribed intervals, so access is limited. These systems are not suitable for high floor loadings and when of timber construction, have limited fire resistance.

Figure 4.4 shows a typical example, comprising 65 mm deep galvanized steel fluted trough decking fixed to the structural floor. The ducts are overlaid with 600 mm wide × 2400 mm long × 22 mm thick flooring grade tongued and grooved chipboard, riveted to the steel troughs and giving an overall raised floor depth of 87 mm. Cables lie in the troughs of the steel decking, with occasional cross-ducts inserted to allow cables to run against the normal direction of the troughs. Access to the floor void is usually provided at this point by 600 mm square removable panels. The system provides built-in electrical screening between power and signal cables. The floor construction can take a maximum uniformly distributed load of 5.0 kN/m², a maximum concentrated load of 2.7 kN applied over an area of 300 × 300 mm, and a partition line load of 2.0 kN/m.

Shallow void battened floors are another development where light loading only is needed and where fire resistance is not required. They are increasingly used where sound insulation needs to be improved floor to floor. Fig. 4.5 illustrates simple batten framework carried on sound insulating pads.

4.4 Deep systems

Deep raised access floor systems are used for very highly serviced areas, such as those requiring air-conditioning plant or computer room cables. The void may be totally accessible by the removal of all panels, or have point access only. The degree of fire resistance relates mainly to the panel material and finish. The supports, termed *pedestals*, allow for adjustability with screw jacks against an uneven structural floor, or one which has been formed at different levels. Floor panels are available which are capable of taking high floor loads, and different lengths of pedestals can be incorporated in the system where there is a need to consider heavily concentrated loads. Except where there are special requirements, such as the need for an antistatic finish in computer rooms, the range of finishes for the panels is limited only by the size of panel.

Figure 4.6 shows a typical example comprising 600 mm

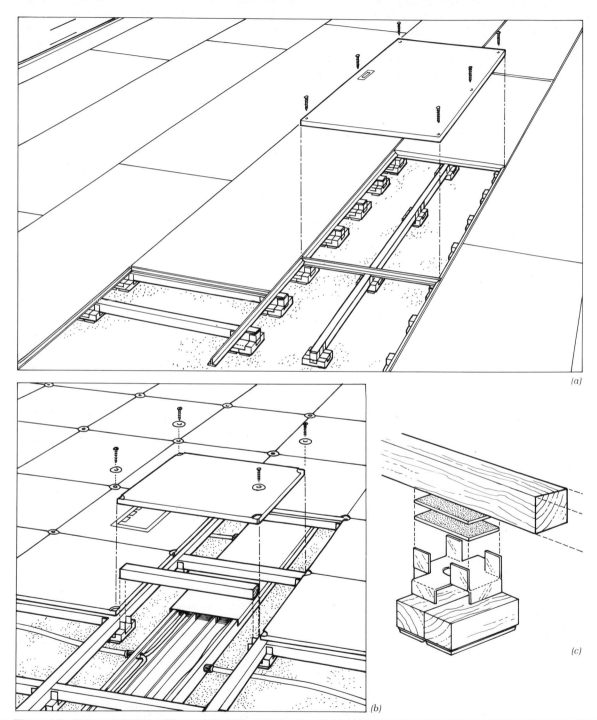

Figure 4.5 Shallow void battened floors (Durabella pattern): (a) partial access; (b) full access; (c) timber and metal cradles.

modular panels of hot dipped galvanized top and bottom trays in which is glued a 30 mm thick high-density particle board. Service outlet boxes can be incorporated in panel cut-outs, but require extra supports to maintain the integrity of the structure. The panels are supported by *stringers*, which in turn are supported by pedestals, fixed to the structural floor.

The stringers are made from zinc-plated steel hollow section, 35 × 35 mm, and are bolted to the specially profiled top plate of the pedestals; accurate location between the stringers is achieved by inserting a four-way gasket. The stringers have sound-deadening gaskets fixed to their topside to isolate the floor panels. Earth-bonding conductors can be incorporated with the stringers and, when required, there can also be a main bonding conductor on the pedestals. For panels with edge trims, an internal electrical connection is made between the top and bottom metal trays; earth-bonding conductors can also be supplied.

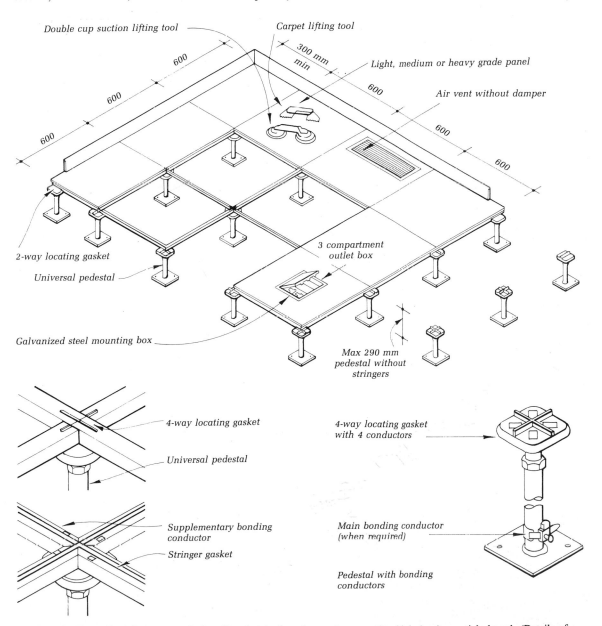

Figure 4.6 Deep raised floor system in hot-dip galvanized steel trays incorporating high-density particle board. (Details of BMA series by kind permission of Donn Products (UK) Ltd.)

The pedestals are of mild steel, coated with a corrosion-resistant finish, and incorporate a sliding tube arrangement with a tightening nut so that the floor can be made level when the horizontal surface of the structural floor is untrue.

The system illustrated complies with the requirements of the PSA's *Performance of Specification: Platform floors* 1985 (Method of Building 01-801), regarding structural performance and deflection test data. Three loading grades of panels are available: *light grade*, maximum uniformly distributed load (UDL) of 2.5 kN/m^2 and maximum concentrated load over an area of 300 × 300 mm of 2.7 kN; *medium grade*, maximum of 5.0 kN/m^2 and 4.5 kN respectively; and for *heavy grade*, a maximum UDL of 12 kN/m^2. The use of particle board floor panels means that the floor has no fire resistance rating, and a class 1 surface spread of flame classification. However, the system satisfies PSA's *Small Scale Fire Test*. Refer also to data available in MOB PF2/SPU dated 1992.

5 Joinery

5.1 Introduction

The following historic textbooks are introduced since background reading is essential to understanding this extensive subject.

J. Eastwick-Field and J. Stillman, *The Design and Practice of Joinery*, Architectural Press Ltd, 1973
C. H. Tack, *Joinery*, PRL, HMSO, 1971
Bill Launchbury, *Handbook of Fixings and Fastenings*, Architectural Press Ltd, 1971

Joinery is generally understood to be the fabrication and fixing of timber components such as windows, doors, stairs, built-in fittings and of external items such as gates, the surfaces of all of which are planed (i.e. *wrot*), and usually sanded. The ease of working timber, and its 'warm' and interesting appearance, encourage its use.

The quality of joinery work depends upon design, materials and workmanship. The designer must, therefore, understand the principles of design, specify exactly the type and quality of timber and other materials, and the standard of workmanship required, and bear in mind the available facilities for manufacture and fixing. The designer should also be aware of the 'green' philosophy regarding the use of tropical timbers from depleted rainforests.

Internally, unprotected wood soon becomes dirty and dull so exposed surfaces are often stained and/or treated with a natural or synthetic resin varnish, or with french or wax polish — clear finishes which considerably enhance the natural appearance of the timber. Alternatively, wood is painted. It is important to note that the smoothness of the wood surface determines that of applied finishes.

All unprotected timbers 'weather' to various shades of grey where exposed to the weather or to constant wetness (sink + bathroom situations). Clear finishes can preserve the new appearance of timber, but they require frequent maintenance. Refer for general information to *MBS*: *Finishes* and to the Timber Research and Development Association (TRADA) for comparative data on clear and stained timber treatments. Timber is often used as a strong and inexpensive core, so that metal-faced plywood, and metal drawn on wood sections, are economical means of obtaining the appearance of metals. Timber windows are available with sections encapsulated in PVC sheet. Increasingly 'solid' timber is being used in conjunction with plywood, blockboard, chipboard, hardboard, plastic laminates and metal sheets and sections. Although sliced wood veneers which often have exotic grain patterns, continue to be glued to surfaces internally, plastics and impregnated paper laminates simulating wood are now commonly used. These can be easily cleaned but they are not resistant to scratching and abrasion and are, therefore, not suitable for counter tops and working surfaces. Principles of good joinery design can be deduced from knowledge of the properties of the materials to be used and the intended conditions of use, e.g. light or heavy duty, internally or externally. Designs can invariably be improved by careful observation of the behaviour of prototypes.

The limitations of hand and of machine work must be taken into account. For example, a spindle cutter cannot form a square end to a groove; either a separate machine operation or handwork is required. Although joinery can be remarkably accurate, inaccuracies are inevitable in fixing, and more so in refixing removable sections such as glazing beads, so these should be either recessed or projected in relation to the joinery to which they are attached (see Fig. 5.1(b)).

5.1.1 Arrises

External corners are better slightly rounded, since sharp arrises are difficult both to obtain and maintain. There is also the tendency for paint and clear finishes to 'run away' from such corners and give early failure. This is undesirable particularly in external situations. The degree of rounding off depends upon the finish chosen. Paints and varnishes benefit from a 3–5 mm radius which has to be machined. Stained or polished hardwoods simply need 'pencil rounding' of 1.5 mm radius that can be obtained by sanding or by use of arris tooling.

The choice of the correct timbers requires knowledge of the characteristics and properties of timbers in general, of

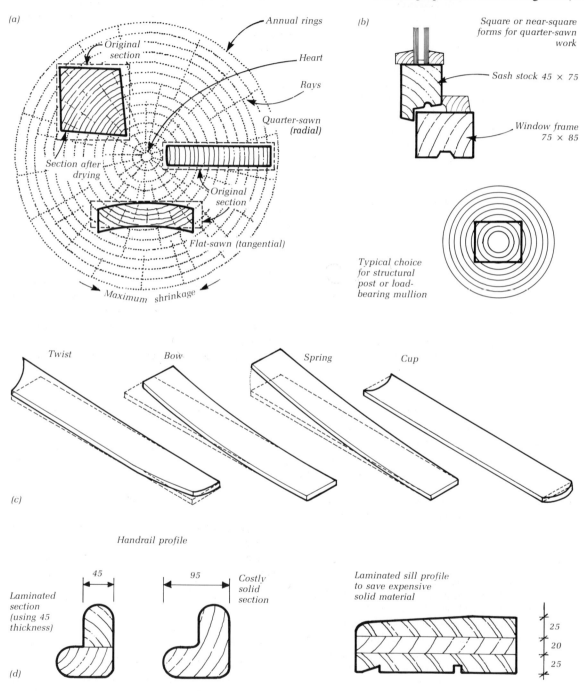

Figure 5.1 (a) Tangential or radial cut components (from *Principles of Use of Materials*, vol. 1, HMSO, 1959); (b) timbers cut square and near square; (c) timbers cut from sawn materials; (d) complex forms build up from laminated sections.

the available species, and of the conditions to which specific joinery components will be subjected (dealt with later in this introduction). Nomenclature, anatomy and properties of timbers, and causes and means of avoidance of deterioration, are dealt with in *MBS: Materials*, Chapter 2.

5.1.2 Species

Traditional naming of timbers is confusing. It will be noted, for example, that the unqualified description *deal* does not relate to any particular species, and the need to use the names given in BS 881 and 589 : 1974 *Nomenclature of commercial timbers, including sources of supply* for hardwoods and softwoods respectively, is emphasized. Designers should ensure that timber from endangered species of trees is not specified. To assist specifiers the Timber Trade Federation maintain up-to-date lists of recommended timber from renewable sources.

5.1.3 Seasoning

All timber for building must be dried slowly (i.e. *seasoned*), if only to avoid too rapid drying and consequent splitting, or to make it receptive to preservatives. A maximum moisture content of 25 per cent is advised for vacuum/pressure impregnation and 22 per cent for organic solvent-type preservatives. Timber for joinery must be dried to levels as near as possible to the relatively low moisture content it will assume, which normally necessitates *kiln seasoning*. This minimizes shrinkage in service, and sometimes expansion, remembering that timber can be too dry. It also reduces thermal conductivity and vulnerability to fungal attack and makes surfaces suitable for gluing and surface finishes. Seasoning is not irreversible, so that priming joinery with paint coats or lacquer 'at works' and protection from the weather in transit to the site and on the site, are necessary. Ideally, timber should not be installed until buildings are heated and 'dried out' with the buildings maintained at constant relative humidities.

5.1.4 Movement

Most joinery must be designed and fixed to permit some moisture movement in service, e.g. by using narrow widths and tongued and grooved joints, or where widths are glued together. Large profile sections for stair strings, mullions or sills can also be built up by laminating timber to avoid the risk of using large scantlings of unsuitable proportion. Another example are benches or 'solid' table tops which employ fixings that allow the overall widths of tops to change with changes in the moisture content of the wood. These changes result from variations in the humidity of the

surrounding air. Paint and clear finishes can only delay such movements.

The extra cost of radially cut timber (see Fig. 5.1(b)) is justified where the smaller movement in its width, and freedom from the 'cupping' of plain sawn timber are critical — as in drawing boards, and/or where an interesting appearance is desired - as that given by 'silver grain' rays in oak.

5.1.5 Thermal and fire resistance

The low thermal conductivity and capacity of timber and the low thermal movement in its length favour its choice for various uses. Timber retains its strength at high temperatures. Sacrificial thickness for charring is allowed in fire-resisting timber frames. A typical historic example is Ann Hathaway's Cottage at Stratford-upon-Avon where a fire destroyed the thatch and infilling to partitions and walls, leaving the original oak frame still standing though reduced by a few millimetres by the flames.

Intumescent strips, which expand in fires, are valuable in sealing gaps around and between fire-resisting doors, and as beads for fire-resisting glazing. The spread of flame classification of timbers can be effectively improved by impregnation with, or by surface applications of, fire retardants, although these cannot make timber non-combustible. There is also the problem that adhesives and surface treatments may not be compatible with retardants. For explanation of terms 'spread of flame classification' or 'non-combustible', see *MBS: Materials*.

5.1.6 BS 1186 Timber workmanship in joinery Part 1

Designers should be familiar with BS 1186 : Parts 1 and 2 which cover the following criteria (for detailed aspects see section 5.3):

- Required moisture content for specific locations (see Table 5.1).
- Straightness of grain and number of growth and number of growth rings per 25 mm.
- Defects — acceptable ones and those that are condemned.
- Patching.
- Standards of adhesives and manufacture for plywood and laminboard.
- A table which advises on suitable timbers to specify for differing joinery components internally and externally (e.g. sills, frames, sashes, external gates).

Changes in timber dimensions will occur despite compliance with BS 1186 and the following additional advice is offered to minimize the problem:

- Longitudinal movement is the least problem for shrinkage. It is wise, however, to choose joinery lengths that fall within economic scantlings. Timbers like pitch pine are available up to 9.000 m in length, but would cause excessive costs in window frame construction; 4.500–6.000 m lengths are the usual maximum sizes for sill or head members.
- Timbers cut tangentially or radially give little movement but are reserved for high-quality joinery (Fig. 5.1(a)).
- Timbers prepared 'flat sawn' and away from heartwood are likely to cup or twist, so that floorboards or linings obtained from that source will need additional fixings (Fig. 5.1(c)).
- Timbers cut to square or near square profile will lead to less problems in use than thin profiles (above 1:2 proportion with flat sawn material) (Fig. 5.1(b)).
- Avoid 'L'-shaped members cut from solid timber and employ glued laminate design to build up shapes or forms that exceed sizes for solid timber (Fig. 5.1(d)).

Although timber expands when it is wetted and returns to its former size when it dries again, it must be remembered that if it is restrained while it absorbs water, when it dries it shrinks from the restrained size and becomes smaller than it would have been had it not been restrained — a phenomenon known as *stress setting*. The effects of this permanent shrinkage are obvious where gaps form between timber flooring after flooding, a more familiar example occurs when wooden tool handles loosen when they dry after having been wet.

5.1.7 External joinery

Unless the timber is inherently durable, it is worth reminding designers that joinery should be designed to minimize the likelihood of wetting by rain or other causes. All such external timber should be protected by damp-proof courses and membranes and flashings, and it should be kept above splash rising from pavings and projecting surfaces. Joinery should be designed to be self-draining with no horizontal surfaces and with devices such as lined channels and weep tubes to remove condensation from glass on the inside of windows. Defective joints, and cracks must be avoided where they would allow water to flow or be blown into, or to enter timber by capillary action. Refer to *MBS: External Components* for a full discussion of these matters.

Precautions are also needed where internal joinery is placed in damp situations such as bathrooms, kitchens, laboratory benches or as draining boards.

A common defect has been the rotting of the lower rails of sashes. Condensation from the inner face of glass enters the wood through defective back putties, the wood swells and cracks the outside paint at a joint, and further water enters (Fig. 5.2(a)). Paint is not a preservative, and in this

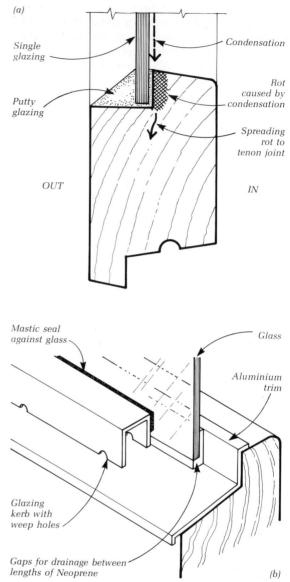

Figure 5.2   (a) Defects due to condensation affecting bottom rails; (b) Norwegian detail to protect bottom rails.

case it only serves to retain water and fungal decay follows, particularly if unpreserved sapwood is present. The sapwood of all species is non-durable and perishable. Sapwood forms a sizeable element in commercial softwoods used today in window production due to the conversion of 'young growth' timbers where logs are 300–450 mm in diameter. It is now considered essential to utilize preservation techniques on all softwoods (due to risk of sapwood) in windows, external doors, frames, trim and cladding, even where the material is painted. There is a good case to be made for recycling joinery prepared from prime grade douglas fir or larch of yesteryear. In the nineteenth and early twentieth centuries this wood was often cut from virgin forest trees. It is of far superior quality in terms of rot resistance compared to the presently available lumber from young forests.

A Norwegian method for protecting the lower members of windows is described in Fig. 5.2(b).

The National House Builders' Registration Council requires timber for claddings, window frames, casements and sashes and external door frames to be treated with preservative, or a preservative and paint system (NHBRC *Practice Note* No. 1 describes acceptable preservative treatment), unless one of the following *durable* timbers is used ('TH' indicates a tropical hardwood):

Afrormosia	
Agba	TH
Afzelia	TH
American white oak	
European oak	
Gurjun	TH
Idigbo	TH
Indian or Burmese Yang	TH
(also known as Keruing	
(Malaysia) or Yang (Thailand))	
Iroko	TH
Kapur	TH
Makore	TH
Red meranti	TH
Red saraya	TH
Sapele	TH
Sweet chestnut	
Teak	TH
Utile	TH
Western red cedar (softwood	
from western Canada)	

It should be noted that many tropical hardwoods are endangered species or may be derived from clear felling of forests. Specifiers having ecological concerns should establish that tropical timbers have been obtained from managed plantations; for example when obtaining teak from the West Indies or gurjun from India or Burma. Home-grown European hardwoods such as chestnut or oak are probably the safest choice. This advice follows discussions with a leading timber importer.

Practice Note No. 1 states that the sapwood of these listed timbers should be treated with preservative. That is an easy solution for western red cedar but impossible to achieve with dense timbers like oak. If the cost of hardwoods can be afforded for sills it is better to specify that sapwood is excluded than risk impregnation that is likely to fail with dense hardwoods.

End grain is especially vulnerable and if it is cut on site the NHBRC advises that ends should be immersed in preservative for at least 1 minute, or if this is not practicable, two brush coats should be applied. It must be noted that preservatives should be applied liberally, unlike paint which is 'brushed out'. Preservatives may adversely affect putties, mastics, window and door furniture and paints. Generally 48 hours must be allowed before applying a primer on a surface which has been treated with preservative and 3—4 days may be necessary where copper naphthenate preservatives are used. Internal joinery made from *perishable* woods, such as beech, do not require preservative if they are kept 'dry'.

Although timber has a high strength : weight ratio it must not be forgotten that some joinery members are highly stressed, if only occasionally, and eventualities such as people standing on tables or balustrading strained on a crowded stair must be foreseen. Concentrations of stress at joints as in those in side-hung, double-glazed casement doors or windows, determine jointing methods and these in turn may determine the minimum sizes of the members to be joined.

The economical use of the *standard sizes* of sawn timber is dealt with in section 5.5.

5.2 Choice of timbers for various uses

If the properties of available timbers are systematically matched to known performance requirements it is often possible to use a timber which is more suited, and yet less costly, than the conventional choice. For example, members in stronger timbers can be smaller and may cost less, than those in 'cheaper' but weaker timbers.

The botanical descriptions *hardwood* and *softwood* are rarely helpful. Thus, although most hardwoods are denser and therefore stronger than most softwoods, hardwoods are not necessarily hard (e.g. balsa) and softwoods are not always soft (e.g. yew). Also, hardwoods include species with heartwood having both the greatest and least resistance to fungal attack, and softwoods include species having both 'small' and 'large' moisture movements.

Timbers vary considerably in their properties and appearance, between species, and even between parts of one tree. Hence, information can relate only to average specimens of any species. In referring to *MBS: Materials* Chapter 2 it is worth recalling that average values are given for the following aspects of timber technology: strength, nailing, gluing and resistance to cutting, blunting effects on tools, drying characteristics, durability and resistance to impregnation and suitability for bending. Designers making their first foray into timber selection for joinery purposes should concentrate their studies on say eight commonly used species, make the effort to know their characteristics and visit timber yards and workshops to become familiar with their choice. That knowledge can be extended to embrace a wider selection of species as work experience increases. A good starting list would be European redwood (*Pinus sylvestris*) and Douglas fir (*Pseudotsuga taxifolia*) representing softwoods and ash (*Fraxinus excelsior*), English oak (*Quercus* spp.), elm (*Ulmus* spp.), iroko (*Chlorophora excelsa*), jarrah (*Eucalyptus marginata*) and Utile (*Entandrophragma utile*) representing familiar hardwoods.

References to official documentation are summarized as follows:

- *Timber Selection by Properties*. Part 1 Windows, doors, cladding and flooring, etc. Princes Risborough Laboratory (PRL) and HMSO.
- *A Handbook of Hardwoods*, PRL and HMSO.
- *A Handbook of Softwoods*, PRL and HMSO.
- *The National House-Builders Registration Council Handbook*. This advice is continually updated and it is important to refer to the current edition when involved with work that incurs NHBRC certification.

The British Standards listed below are the most useful:

BS 373 : 1957 *Testing small clear specimens of timber*.

BS 1186 : 1957 *Timber and workmanship in joinery* Part 1 : 1986 *Specification for timber* Part 2 : 1971 *Quality of workmanship*.

BS 4978 : 1973 *Timber grades for structural use*.

BS 5268 *Code of practice in the structural use of timber* Part 2 : 1984 *Code of practice for permissible stress design, materials and workmanship* Part 4 *Fire resistance of timber structures*.

BS 5756 : 1980 *Specification for tropical hardwoods graded by structural use*.

BS 5820 : 1979 *Methods of test for determination of certain physical and mechanical properties in structural size*.

BS 6100 *Glossary of building and civil engineering terms* Part 4 *Forest products* Section 4.2 : 1984 Sizes and qualities of solid timber.

BS 1186 : Part 1 gives the suitability of 36 hardwoods and 10 softwoods which are available in this country, for 12 joinery applications.

5.3 Joinery specification

The variation in timber quality and factors like varying shrinkage means that phrases such as 'joinery to be *free* from all defects' are not sustainable. Specifications refer to joinery as 'reasonably free' from defects and this certainly needs more accurate definition to avoid arguments.

Guidance is available from a range of British Standards, the following having a crucial role to play:

BS 459 *Doors* Part 4 : 1988 *Matchboarded doors*.

BS 1576 : 1953 Wood door frames and linings.

BS 4471 : 1987 *Specification for sizes of sawn and processed softwood*.

BS 1186 *Timber and workmanship in joinery* (this is a key British Standard and is referred to as follows and in section 4.1): Part 1 : 1986 *Specification for timber* Part 2 : 1971 *Quality of workmanship*.

BS 1186 : Part 2 : 1971 details quality requirements for timber in four 'use classes', i.e. Class 1S — joinery for clear finishing and Classes 1, 2 and 3 for painting. Rules are given for 'concealed' and 'semi-concealed' surfaces. The Standard must be consulted for details, but the following notes will give guidance:

1. Timber must be free from fungal decay, and from insect damage other than pinhole borer (ambrosia) holes which are permitted in concealed and semi-concealed surfaces, and if the holes are filled, also in Class 1, 2 and 3 surfaces. It is advisable to exclude pinhole borer in carcase work where expensive veneering is proposed.
2. Sapwood is not allowed in hardwood surfaces exposed to the weather, or in wet surroundings within a building (a better specification for high-class work would exclude sapwood altogether).
3. Unsound, dead and loose knots are restricted to concealed and semi-concealed surfaces and knot sizes are limited.
4. Laminating, finger jointing and edge jointing must not be unduly conspicuous and may be disallowed by the purchaser for Class 1S use. For this use also, the species and character of grain must be the same on all surfaces and be matched as far as possible.
5. For Class 1 timber, checks and shakes are restricted in size and depth. Not less than eight growth rings per 25 mm are specified and the slope of grain is restricted to not more than one in eight in hardwoods and one in ten in softwoods.
6. Sapwood (except that mentioned in item 2) including discoloured sapwood, is allowed. Once again a better specification for softwood joinery would exclude sapwood entirely.

The recommended moisture contents for all classes of joinery are given in Table 5.1.

5.4 Joinery workmanship

The traditional joiner or carpenter had great pride in providing a high standard of workmanship. The phrase 'the work is to be performed in a workmanlike manner', needed no provisos. Jon Broome's account in 'The Self-build Book' shows that this spirit has not disappeared. Tracy Kidder's

Table 5.1 Moisture contents for joinery when handed over to the purchaser

Joinery	Moisture content percent ± 2
External	
Heated or unheated buildings	17
Internal	
Heated buildings	
(i) intermittent	15
(ii) continuous, with room	
temperatures of 12−18°C	12
20−24°C	10
(iii) timber in close proximity to	
sources of heat	8

fictionalized version of carpenters working in Massachusetts sounds remarkably like my son's experience of house construction in Virginia.[1] All those worlds are removed from building under JCT or other forms of contract where precise descriptions are necessary. BS 1186 : Part 2 : 1976 is once more the key reference with a section titled 'Quality of Workmanship in Joinery' and which is highlighted as follows:

- Requirements are stated for fit of parts and care in forming appropriate joints, including glue work (this does not excuse the designer from designating joints). Framing is described and standards stated for flatness and squareness.
- Tolerances between components such as painted and unpainted doors and sashes and fit of drawers.
- Construction of laminated and finger joints to enable short lengths to be joined together.
- Protection of joinery from exposure and from damage in transport, on site and after installation.
- Definitions as to finish, namely for painting, or to a higher standard for staining or polishing.

It should also be remembered that the initial British Standards were drawn up to arrive at acceptable standards for post-war housing in the 1940s and 1950s. In joinery a minimal British Standards attitude still exists as compared with Scandinavian practice. Work made to higher performance standards of Norway, Sweden or Denmark is available from licensed manufacturers in the UK apart from imported joinery. One of the pleasures of working on smaller contracts is the working relationship that can be built up between designer and joinery specialist where over the years a mutual skill is developed in joinery detailing. The same situation is feasible on larger works provided specialists are nominated for high quality woodworking (see section 5.15).

Protection of quality joinery is important where specific directions are needed. First and foremost, the heat and humidity conditions in the building must be suitable for joinery to be delivered and fixed. This means ensuring that the environment is commensurate with the required moisture content of the timber (see Table 5.1). Clear seals and film are available which can be applied to joinery in the workshop to give additional protection in transportation and on site before the final finishes are applied. Protection on site often includes boxing in with hardboard or plywood. Staircases are best installed after plastering. Highly finished door linings or cupboard fronts are also fixed at a later stage over wrot grounds or battens that are set in place at the plastering stage. The use of carefully prepared wrot grounds means that architraves and skirtings can be fixed by gluing instead of employing pins or screws. In nineteenth-century London it was common practice for residential properties to be painted out 'in white' (undercoat and distemper) and to be let for 12-months prior to disposal by lease. This enabled the elaborate final decorations to be delayed until joinery and plaster defects were put right — an idea that might still be relevant in restoring historic buildings.

Many elements in joinery play a structural role such as loadbearing mullions or muntins in windows and glazed screens. Partition and shelving systems perform as structures, as do timber-framed stairs and balustrades. It is necessary to employ calculations or to demonstrate by testing that the assembled elements will perform under stress. Structural engineers are now associated with the design of stressed joinery, the engineering input coming via consultants or through the design services of specialist manufacturers. Traditional pattern books of construction gave notional sizes for joinery profiles, today's Codes of Practice, performance based specification and stress grading imply a more scientific approach and particularly where reliance is placed on modern adhesives, often far stronger than the timber that is being joined together (Fig. 5.14).

Sizes for softwood joinery should be based upon standard sizes (see Table 5.2 and Figs 5.3 and 5.4); hardwoods offer greater freedom since joinery works may be converting baulks to the sizes needed for the work in hand. However, timber suppliers that serve the building trade market a whole range of prepared softwood and hardwood trim that will be cheaper to purchase than specially milled profiles.

The standard trim referred to includes stock sections for casements and frames as well as architraves and skirtings. The former is targeted at simple refurbishment work and will not meet the requirements for 'high performance' joinery where higher standards of sound and/or thermal insulation are relevant. Makers of high performance joinery maintain stocks of their standard profiles. A compromise solution in custom-made joinery is to incorporate manufacturers' standardized details made to special sizes.

If the budget permits sections milled to purpose-made designs, then the number of sections should be limited to

Table 5.2 Reductions from basic sizes to finished sizes to accurate sizes by processing of two opposed faces of softwoods

Purpose	Reduction from basic size to finished size for sawn width and/or thickness (mm)				
	15–22	22–35	35–100	100–150	Over 150
Construction timber, surfaced	3	3	3	5	6
Floorings*	3	4	4	6	6
Matchings and interlocking boards*	4	4	4	6	6
Planed all round Trim	5	5	7	7	9
Joinery and cabinet work	7	7	9	11	13

* The reduction of width is overall the extreme size exclusive of any reduction of the face by the machining of a tongue or lap joint.

the absolute minimum. The high cost of custom-made work rests with the expenditure in the setting up of cutters, planers and routing machines; this initial work needs to be recovered by the repetition work on a very large order for joinery production. Profiles should not only be drawn to full size but dimensioned. It is worth recalling that many joinery works in the UK operate with imperial sized machines so that metric sized sections of timber are often rebated or grooved to imperial dimensions.

One of the unusual aspects of the Usonian homes by Frank Lloyd Wright was the reduction of all joinery trim to half a dozen or so profiles. Those profiles performed the role of casement or door rails, frames, architraves, cover battens, dados and skirtings throughout the whole construction.[2]

5.5 Sizes of softwoods

Metric sizes have been agreed by all the major softwood-

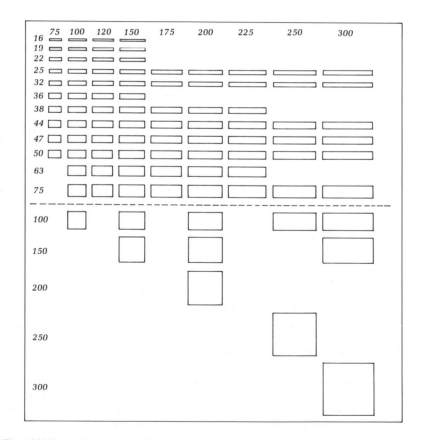

Figure 5.3 BS 4471 : 1986 Basic cross-sectional sizes of sawn softwood at 20 per cent moisture content. The smaller sizes contained within the dotted line are normally, but not exclusively, European. The larger sizes outside the dotted lines are normally, but not exclusively, of North and South American origin.

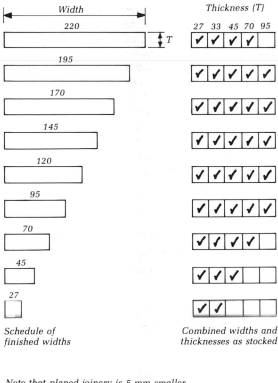

Schedule of finished widths

Combined widths and thicknesses as stocked

Note that planed joinery is 5 mm smaller than nominal sizes sold, e.g. 220 × 70 is prepared from 225 × 75

Figure 5.4 Timber merchants' schedule. Typical schedule giving standard finished sizes for planed softwoods. (By kind permission of A. W. Champion Ltd.)

Table 5.3 Finished widths and thicknesses of small resawn softwood sections

Finished thickness (mm)	Finished widths (mm)				
	22	30	36	44	48
6	X		X		
14			X		X
17	X		X		X
22	X	X	X	X	X
30			X	X	X
36			X		X
44					X
48					X

Table 5.4 Basic cross-sectional sizes of sawn heartwoods having 15% moisture content

Thickness (mm)	Width (mm)			
	50, 63	73	100, 125	100, 175	200, 225 250, 300
19		X	X	X	
25	X	X	X	X	X
32		X	X	X	X
38		X	X	X	X
50			X	X	X
63				X	X
75				X	X
100					X

producing countries and the principal European importing countries. Figure 5.3 shows the relevant data from BS 4471 : Part 1: with sawn supplies minus 0.5 mm permitted off not more than 10 per cent of the pieces in any parcel and plus 3 mm on any proportion of a parcel. Where imperial sized lumber is needed to match existing buildings, sources are still available from Canada and the USA.

Minus deviations permitted on up to 10 per cent of the pieces in any sample are 1 mm on widths and thicknesses up to 100 mm and 2 mm on greater sizes.

The Standard provides for *precision timber* produced by machining *(regularizing)* at least one face and edge of a section to give a uniform thickness and/or width throughout 1 mm less than the basic sawn size. This quality of precision timber is used in stick-framed buildings, and for loadbearing partitions and trusses. It will allow for more accurate framing and provide level fixing surfaces.

Table 5.2 gives the reductions allowed for planing sawn sections to accurate sizes, ranging in the case of joinery from 7 to 13 mm, according to sizes of pieces. Plus or minus 0.5 mm is allowed on all finished sizes. Lengths of softwoods are from 1.8 to 6.3 m rising by 300 mm increments. It is more practical to work from the scheduled finished sizes offered by timber merchants as given in Fig. 5.4 and Table 5.3.

5.6 Sizes of hardwoods

BS 5450 : 1977 *Specification for sizes of hardwoods and methods of measurement* gives sizes of hardwoods and methods of measurement — as Table 5.4. Hardwoods are not necessarily imported in British Standard sizes. Therefore availability in any particular species should be checked. Permitted deviations from the basic sizes are given in Table 5.5.

Sizes will be similar for moisture contents less than 15 per cent and greater for moisture contents up to 30 per cent to an extent which can be estimated using the values for radial and tangential movements given in *The Handbook of Hardwoods*. Hardwood is stocked in one of the specified thicknesses and is often in random widths with either square or waney edges.

Table 5.6 gives the reduction allowed for planing sawn

Table 5.5 Permissible deviations from basic thicknesses or widths or hardwoods at 15% moisture content

Basic size (mm)	Minus (mm)	Plus (mm)
Under 25	1	3
25–75	2	6
76–125	3	9
126–300	4	12

Table 5.6 Reductions from basic sawn sizes to finished sizes by processing two opposed faces of hardwoods

End use or product	Reduction from basic size to finished size for basic sawn sizes of width or thickness (mm)				
	15–25	26–50	51–100	101–150	151–300
Constructional timber, surfaced	3	3	3	5	6
Floorings, matchings and interlocked boarding, planed all round	5	6	7	7	7
Trim	6	7	8	9	10
Joinery and cabinetwork	7	9	10	12	14

sections for accurate sizes, ranging in the case of joinery from 7 to 14 mm according to the sizes of the pieces. Finished sizes after processing are allowed ± 0.5 mm deviation. A stricter rule could be applied whereby profiles are to be as dimensioned with a positive deviation only of 0.5 mm, this permitting sanding down in assembly to give a smooth face. Detailing in hardwood means sizing components to precise finished dimensions.

Lengths of imported hardwoods vary according to species and origin. British Standard *basic lengths* are any integral multiple of 100 mm but not less than 1 m.

5.7 Building boards

Plywood, blockboard, laminboards, densified laminated wood, particle boards, fibre building boards and other boards which may be used in the manufacture of components are discussed in *MBS: Materials*, Chapter 3. Over recent years the use of medium-density fibreboard (MDF) has gained importance in use for the manufacture of furniture. BS 1105 : *Fibre building boards* Part 2 : 1971 *Medium board and hardboard* was amended in 1985 for its inclusion. The main characteristics of MDF are homogeneity and ease and precision of machining. Both faces are smooth and the density is of the same order as the high density medium hardboard. Essentially, MDF is a high-quality interior board for furniture, decorative panelling and similar manufacturing purposes, moisture-resistant boards are available which extend its uses for bathroom and kitchen fittings. The limiting factor is the commonly stocked size of 2.4 × 1.2 m. Longer lengths up to 3.0 m can be obtained from a few suppliers.

5.8 Joints

The designer needs to refer again to BS 1186 *Quality of timber and workmanship in joinery*. Part 2 : 1971 *Quality of workmanship* where detailed requirements concerning fit, tolerances and general workmanship are numerated and qualified for the more common joints in woodworking. The choice depends upon the relationships and shapes of members to be joined. Joints may need to be designed to relate to rebates and mouldings in the members (Fig. 5.5). Other factors include strength and appearance (including 'secret' methods) (see Fig. 5.10). There may be the need for demounting (see Figs 5.13 and 5.20). Another aspect is the ease of making and the final cost.

In recent years, the smaller sizes of many members and the greater cost of forming framed joints have led to increased use of mechanical methods for jointing in cabinet work. By 'mechanical' one refers to metal coupling bolts or angle cleats which allow joinery to be screwed together.

Strength requirements for joints vary considerably, from those which locate members during construction only, to those which are heavily stressed in service in compression, tension, shear and/or torsion, and in one or more directions. The strength of an unglued interlocking joint, such as a mortice and tenon, is related to the reduced sectional area of the weaker member. Nails, screws, bolts and dowels have the disadvantage of concentrating stresses in a limited number of small areas, whereas adhesives, which are generally stronger than the wood they join, distribute stresses more evenly. Combinations of glue and mechanical forms may be beneficial in service, or in assembly where interlocking shapes or inserts hold members together while glue is setting.

The guidance in Table 5.7 suggests examples of types of joints appropriate for stated typical situations. The types are listed in the table, and described under the following headings:

- interlocking, e.g. tongued and grooved
- inserts, e.g. dowels, nails and screws
- demountable connectors
- adhesives

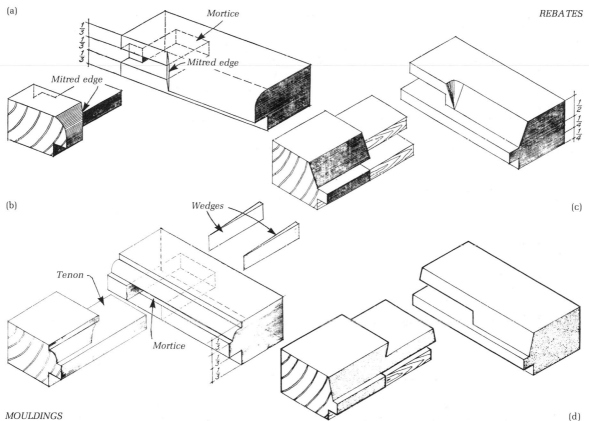

MOULDINGS

Figure 5.5 Joints in relation to mouldings and rebates: (a) mitred moulding; (b) scribed moulding; (c) combed joint with splay; (d) combed joint with splay formed with a router.

5.9 Interlocking joints

5.9.1 Mortice and tenon (Fig. 5.6(a)–(j))

This is the most common means of joining 'flat' rectangular sections of joinery at right angles. The mortice is a slot cut in (usually through) one of the members to receive a tenon projecting from the end of the other member. The tenon is glued in the mortice, and in handwork, it is wedged. The tenon and wedges project initially, but when the glue has set the surplus timber is cut off to give a flush face. For mass production, instead of wedges, the joint is secured by non-ferrous metal star-shaped dowels driven from the face through the joint. This, of course, restricts the finishing work which can be done later and hardwood dowels are better in this respect.

The thickness of a tenon should not be more than one-third of that of the section, and its depth should not exceed five times the thickness. Tenon joints are shown in Fig. 5.6(a)–(j) as follows:

(a) *Through tenon*   In its simplest form this joint is used for joining intermediate rails and stiles in doors. The top and bottom edges of the mortice should be cut so that the slot is slightly dovetailed, thus increasing the strength of the joint. This dovetailing effect is obtained by moving the timber slightly from side to side during the machine cutting process. Joints may be additionally secured by dowels, particularly in large sections of timber.

(b) *Haunched tenon*   This joint is used to connect the stile and top rail of a door, since in order to wedge the joint it is necessary to retain a thickness of timber above the tenon. The cutting away of the front part of the tenon at the top does this, while the retention of the haunch minimizes any loss of strength. In making the joint, the stile is cut so a 'horn' projects beyond the top rail in order to resist the pressure from the wedges when the joint is made. The 'horn' is then cut off level with the top of the door and the wedges trimmed to size.

Table 5.7 Common joints for typical joinery applications (figure numbers)

Typical joint applications	Mortice and tenon	Tongued and grooved	Housed	Combed	Dovetailed	Lapped	Mitred and/or scribed	Finger	Dowelled	Loose tongue	Bolts	Demountable connectors	Adhesives
										Inserts			
'Secret' joints				5.6 (k)							5.18 5.19		
Demountable joints												5.20	
Movement joints									5.15 (a)(b)				
'Thick' sections e.g. window frames Window sills: angles									5.15 (a)(b)				
Handrail and sills: end — end										5.19			
Curved — straight	5.6 (j)									5.19			
Post — sill	5.31 (b)(c)												
'Flat' sections, e.g. boards Architraves — angles													
Moulded sections			5.5 (c)(d)				5.5 (a)(b)						
End — end								5.13					
Edge - edge									5.14 (a)(b) 5.15 (a)(b)		5.21		
Framing	5.7 (a)—(f) 5.8 (f)—(j) 5.9 (b)				5.11 (d)—(f)						5.21		
Door lining: head — jamb	5.6(d) (f)—(i)				5.11 (a)—(c)								
Drawer: side — front/back				5.9 (a) 5.10 (a)—(c)	5.12 (a)—(d)								
Shelf — upright			5.8 (a)—(c)	5.8 (c)—(e)					5.14 (a)				
Stile: intermediate rail of panelled door	5.6 (a)—(d)						5.5 (a)—(d)		5.6 (a)(c)				
Stile: middle rail of panelled door	5.6 (c)—(e)						5.5 (a)—(d)		5.6 (a)(c)				
Stile: top rail of panelled door	5.6 (b)						5.5 (a)—(d)		5.6 (a)(c)				

Note: Interlocks may require to be supplemented by inserts and/or adhesives, but inserts, demountable connectors and adhesives may be the sole method of fixing.

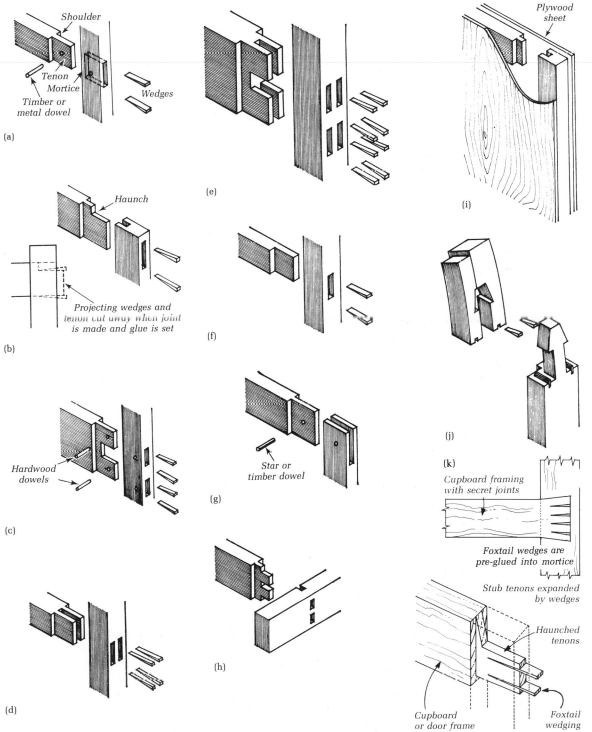

Figure 5.6 Mortice and tenon joints: (a) through tenon; (b) haunched tenon; (c) twin tenon; (d) double tenon; (e) twin double tenon; (f) bare faced tenon; (g) open or slot mortice tenon; (h) twin double haunched tenon; (i) stub tenon; (j) hammer head key tenon; (k) fox-tail tenon.

(c) *Twin tenon* Where the mortice and tenon joint is to be made in a deep rail, say 230 mm and over, there would be a tendency for a single deep tenon to shrink and become loose. To avoid this, two tenons are cut one above the other out of the depth of the rail. In good class work the joint is dowelled as well as being wedged.

(d) *Double tenon* Where a rail is more than, say, 65 mm thick two tenons are cut side by side.

(e) *Twin double* A combination of twin and double tenons is used where the rail is deep and over 50 mm thick. Double and twin tenons can be haunched by leaving shoulders of timber at the top and bottom of tenons. This joint allows a mortice lock to be fitted with less weakening of the framework.

Confusion must be guarded against in naming tenons. For example, twin tenons are sometimes called a pair of single tenons and the terms 'twin' and 'double' are interchanged!

(f) *Bare faced tenon* This variation, which is used when the two members to be connected are of different thickness, allows one face of the work to be flush.

(g) *Open or slot mortice* This joint is easily made. The tenon or tenons cannot be wedged and are secured by gluing and dowelling. It is often used where the framework will be concealed.

(h) *Twin double-haunched tenon* This is a locating joint only, and therefore it is not wedged.

(i) *Stub tenon* Like the open mortice joint this joint cannot be wedged, but is sufficiently strong where both sides of the frame are strengthened by plywood or hardboard.

(j) *Hammer head key tenon* This very strong joint is used to connect curved members to uprights.

(k) *Fox-tail tenon* This complex form is used in high-class handwork and is a 'secret' joint, wedges being (permanently) incorporated in the joint during assembly.

5.9.2 Tongued and grooved (Fig. 5.7(a)–(f))

This joint is used principally for joining boards edge to edge, a tongue cut on the edge of one board fitting into a groove cut into the edge of the other. Without glue the joint allows boards to move in their width. With glue the joint locates boards while the glue is setting, although at the cost of extra width equal to the extent of penetration of tongues into grooves. Grooves are often cut slightly deeper than the projection of tongues to ensure a tight joint on the face. Where the reverse face will not be seen and where boards are not to be glued and cramped, the back shoulder can also be cut to remain slightly open.

Figure 5.7(a) shows a *square edge tongue* characteristic of handwork and (b) and (c) show splayed and rounded tongues which are more suitable for machine work. Figure 5.7(d) is a *Lindermann* machine-made joint with an offset dovetail; it has been used for the strong joints needed for forming deep stair strings. Figure 5.7(e) and (f) shows *square and splayed double tongued and grooved machine-made joints* which give a larger gluing surface and are suitable for joining boards of 40 mm and more in thickness.

5.9.3 Housing joints (Fig. 5.8(a)–(j))

(a) *Square housing* A straightforward method of locating two pieces of timber being jointed at right angles as in shelving. The joint requires careful machining and gluing and/or fixing with screws or nails, punched in and filled. Other examples of tongued and grooved joints are lapped and rebated corner joints.

(b) *Shouldered housing* The groove here is less than the thickness of the horizontal member; it will require additional fixing throughout the face of the vertical member.

(c) *Stopped housing* The groove is stopped back from the face of the upright. This is done where the improvement in appearance is considered to justify the increase in cost.

(d) *Dovetail housing — single* The groove is cut square on one side and given an upward chamfer to form a single dovetail on the other. The key profile so formed helps to prevent the joint pulling out. This joint can only be assembled by sliding the two parts together, and so the joint is not suitable where the pieces to be connected together are wider than say 300 mm.

(e) *Dovetail housing — double* The groove is fully dovetailed with a corresponding increase in strength as against the single dovetail housing. There is, of course, a corresponding increase in cost and difficulty of assembly. Additional fixing by screwing or pinning is not so necessary.

(f) *Rail housing* A form of stopped housing used in framing between rails and end pieces where the rail is not as wide as the end piece.

(g) *Double stopped housing* Used in skeleton framing where a neat appearance is required; this joint is principally a *locating* joint not having the strength of the other type of housing.

(h) *Face housing* This easily made joint is used in skeleton framing which is subsequently concealed. In these positions the fixing can be by nails or screws.

(i) *Shouldered face housing* This modification of the simple-faced housed joint provides additional area of contact while reducing the amount of timber to be cut away from the vertical member.

(j) *Dovetailed face housing* A housed joint incorporating the strength of the dovetail profile.

Another common housed joint is that used to join stair treads and risers to strings. In this case, the underside of the treads is usually concealed, and the housings are tapered to receive wedges.

5.9.4 Combed joints (Fig. 5.9(a) and (b))

These are simple machine-cut joints which have a larger gluing surface than butt joints. They should have a push fit. In thicker sections the number of tongues and slots can be increased with advantage.

5.9.5 Dovetailed (Fig. 5.10(a)–(c))

The parts of these joints are cut to give a mechanical 'lock' in one direction. Figure 5.10(a) is a hand-made *common dovetail* used in high-quality joinery, particularly in joints between the sides and backs of drawers, and (b) shows a hand-made *stopped (and lapped) dovetail* joint suitable for joints between the sides and front of drawers. It is hand-made and would be used for high-quality work only. In the typical version shown in Fig. 5.10(c), a special machine cuts tails and sockets together in timber up to about 225 mm wide. All machine-cut dovetails are stopped and lapped.

Other examples of dovetails shown in this chapter are the hammer head key joint, the Lindermann joint, square and face housing joints and a mitre with secret dovetail.

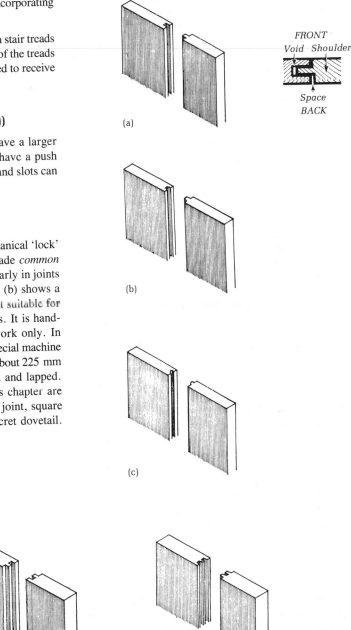

Figure 5.7 Tongued and grooved joints: (a) tongued and grooved — square edge tongue; (b) tongued and grooved — splayed tongue; (c) tongued and grooved — rounded tongue; (d) Lindermann; (e) double tongued and grooved — square tongue; (f) double tongued and grooved — splayed tongue.

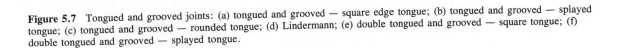

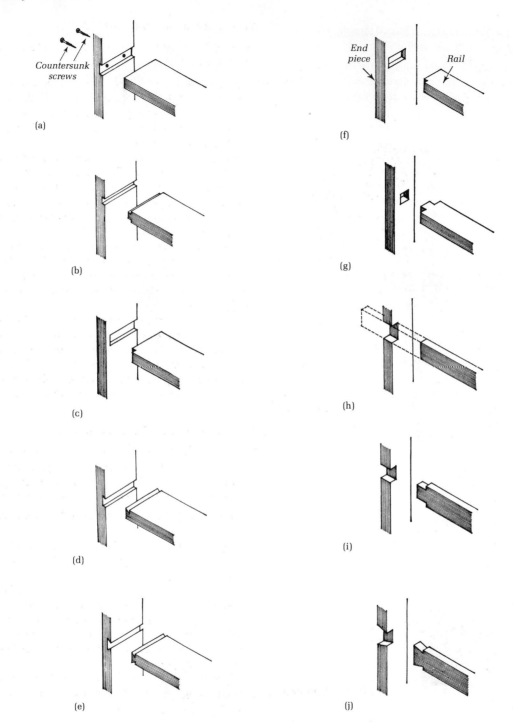

Figure 5.8 Housing joints: (a) square housing; (b) shouldered housing; (c) stopped housing; (d) dovetail housing — single; (e) dovetail housing — double; (f) rail housing; (g) double stopped housing; (h) face housing; (i) shouldered face housing; (j) dovetailed face housing.

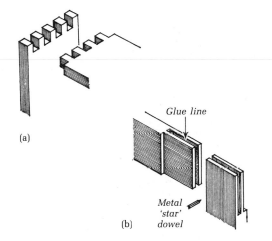

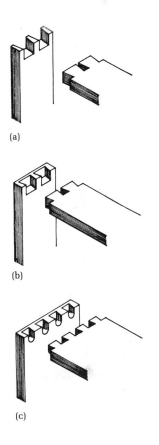

(a)

Glue line

Metal
'star'
dowel

(b)

Figure 5.9 Combed joints: (a) corner locking joint; (b) tongued and doweled joint.

(a)

(b)

(c)

Figure 5.10 Dovetailed joints: (a) common dovetail; (b) stopped or lapped handwrot dovetail; (c) stopped and lapped — machine-made dovetail.

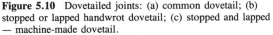

5.9.6 Lapped joints. (Fig. 5.11(a)–(f))

This principle is used in various forms, including the stopped dovetail joints:

(a) *Rebated* A lapped joint which allows the joining of two pieces of timber at right angles and at the same time conceals the end grain on one face. As with all lapped conditions the joint is secured by pinning or screwing and gluing.

(b) *Lapped and tongued* A joint commonly used between the head and jamb of a door lining, it is stronger than the rebated joint but does not conceal the end grain.

(c) *Rebated and tongued* This joint is not much used, although it combines the merits of rebated and lapped and tongued joints.

(d) *Half lapped* This is the most usual way of extending the lengths of members where the joint is fully supported. The overlap should be screwed and glued. Other lapped joints include stopped dovetails and lapped mitres.

(e) *Notched joints* These simple joints provide single or double interlocks between lapped rectangular sections.

(f) *Cogged joints* These are similar to notched joints and are used in the same circumstances. For joinery, notched and cogged joints would normally be glued while for carpentry they would probably be nailed.

5.9.7 Mitred joints. (Fig. 5.12(a)–(d))

Mitres conceal end grain, the grain runs continuously around the exposed faces and the appearance is symmetrical. They are used to connect boards at their edges. Mitred joints are shown in Fig. 5.12(a)–(d) as follows:

(a) *Plain* A mitred joint strengthened with a square block screwed to the boards.

(b) *Lapped* A mitred joint which gives a greater gluing area for increased strength. The shoulders ensure an angle of 90 degrees.

(c) *Mitre with loose tongue* The tongue, preferably of plywood, locates and strengthens the joint. The groove can be stopped so the tongue is not seen on the surface, although this involves handwork. To ensure a tight fit, the tongue is slightly narrower than the combined width of the grooves.

(d) *Mitre with secret dovetail* This mitred joint has the added strength provided by a secret dovetail, but it cannot be cut by machine.

5.9.8 Finger jointing

The finger joint is a strong end-to-end joint which can reduce waste of costly timber by joining short pieces

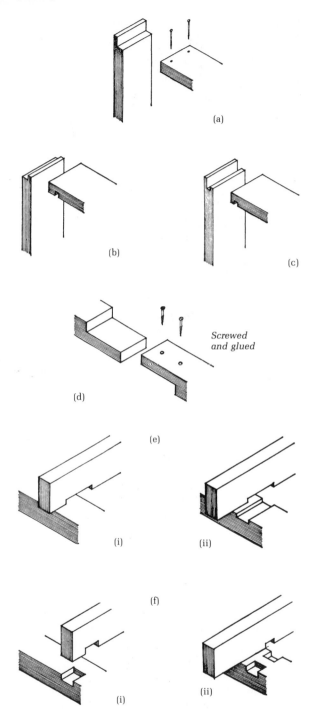

together (Fig. 5.13). The proportions of the fingers vary according to the stresses expected. Refer to BS 1186 : Part 2 for technical aspects.

5.10 Inserts to joints

These joints employ wood or metal connecting devices such as dowels, loose tongues, nails, screws and bolts either as primary fixings, to reinforce interlocking or glued joints, or to retain members while glue is setting.

5.10.1 Dowelled (Fig. 5.14(a) and (b))

Holes for dowels do not weaken timber sections as much as mortices, and properly glued hardwood dowels provide sound joints which usefully reduce the effective length of one member in a joint. The necessary true alignment of dowels and holes is difficult to achieve by hand.

Dowels, with grooves which allow surplus glue to escape, should project about 25 mm, and be a 'push fit' into holes which are very slightly larger in diameter and about 4 mm deeper than dowels, so the shoulders of the joint fit tightly. To help fitting joints, the ends of dowels can be chamfered, although this involves handwork. Shoulders of dowelled joints should also be adequately glued. Figure 5.14(a) and (b) shows typical dowelled joints. The dowels are at about 150 mm centres along the edge−edge joint.

5.10.2 Loose tongue (Fig. 5.15(a) and (b))

(a) *Loose tongue*   Both pieces of timber forming the joint are grooved and a so-called hardwood or preferably a plywood 'loose' tongue is glued in place on one side of the joint to strengthen it. The whole of the joint is then glued and 'cramped up' until the glue is set. This joint is commonly used in table and bench tops.

(b) *Loose tongue with dowels*   This shows a joint between two members which are sufficiently thick for two rows of dowels as well as a plywood loose tongue. The latter is more easily inserted after the joint has been assembled.

5.11 Nails and pins

The basic types are shown in Fig. 5.16; for further details refer to BS 1202 and to makers' catalogues and timber engineering leaflets.[3] The metals used for nails are steel, stainless steel, copper and duralumin. The finishes available for steel are as follows:

- Bright, that is polished by cleaning in rotating bins (rusts in presence of moisture).

Figure 5.11 Lapped joints: (a) rebated joint; (b) lapped and tongued joint; (c) rebated and tongued joint; (d) half-lapped; (e) notched joints: (i) single, (ii) double; (f) cogged joints: (i) single, (ii) double.

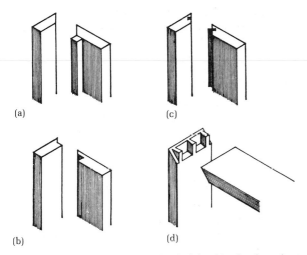

(a)

(c)

(b)

(d)

Figure 5.12 Mitre joints: (a) mitred plain; (b) mitre lapped; (c) mitre with loose tongue; (d) mitred dovetail.

l – *Finger length*
t = *Distance between fingers*
b= *Width of the finger tip*

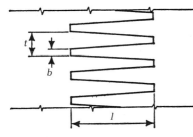

Figure 5.13 Finger joints: l = finger length; t = distance between fingers; b = width of fingertip.

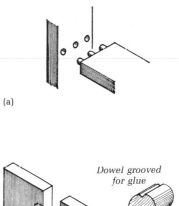

(a)

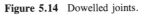

Dowel grooved for glue

(b)

Figure 5.14 Dowelled joints.

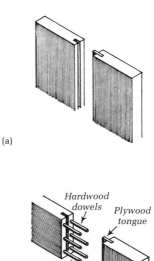

(a)

Hardwood dowels   *Plywood tongue*

Grooves for gluing

(b)

Figure 5.15 Loose tongue joints: (a) loose tongued; (b) loose tongue with dowels.

- Coated, namely molten zinc (termed 'galvanized' and employed externally and where damp situations exist). Other coatings include plastic and shellac (termed 'japanned'). The latter are both decorative but not durable.
- Electroplated finishes, namely brass, cadmium, copper, nickel, tin or zinc. All have decorative qualities and provide some protection from rusting; zinc plating is the best for durability.

The majority of nails today are made from drawn wire, those cut from sheet are termed 'cut clasp' or 'cut floor brads'. The most common in Fig. 5.16 are described by method of framing, type of head and shank, base material and finishes.

Other characteristics of nails are as follows:

Maximum and minimum lengths given to nearest mm

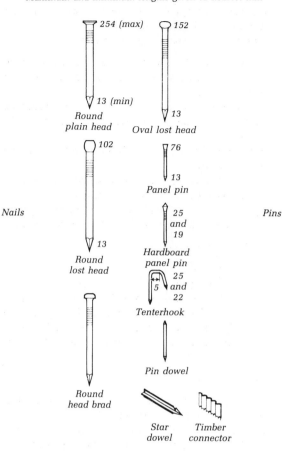

Nails

Pins

Figure 5.16 Nails and pins.

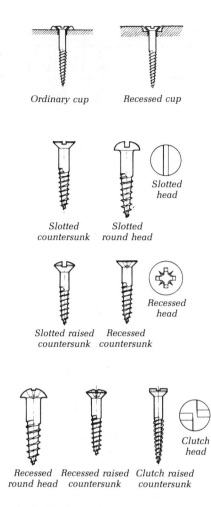

Figure 5.17 Wood screws and cups.

- *Lost-head nails* have small heads so they can be punched below surfaces.
- *Improved nails* including *annular ring* and *helically threaded* types, resist withdrawal better than round wire nails.
- *Automatic nails* are provided in strips or coils for use in pneumatic guns or manual nailing machines. They are often resin coated to improve withdrawal resistance.
- *Nail heads*, in timber which is to be painted or to receive clear treatment, are usually punched below the surface and the hole is filled with proprietary filler. Where a surface is to have a clear finish the filler should be of matching colour.

Externally, steel nails should be zinc coated, but they should be well punched in, as the coating may be damaged.

- *Panel pins* are used for purposes such as securing sheet materials until non-contact adhesives have set.
- *Timber connectors* are often called *corrugated wedge fasteners*, and are used to reinforce joints.

5.12 Wood screws

Screws are drawn from wire, and their materials and finishes are similar to those described for nails except that brass replaces copper. BS 1210 gives information on slot head screws and some patterns of recessed head, and CP 112 specifies centres for fixings but these should be checked against potentially more up-to-date information available from manufacturers of sheet materials. Table 5.8 gives screw gauges.

Most finishes on steel screws are decorative rather than protective. Stainless or non-ferrous types should be used for external locations where good durability is required

Table 5.8 Screw gauges

Screw gauge	Nominal diameter of screw and unthreaded shank (mm)
0	1.52
1	1.78
2	2.08
3	2.39
4	2.74
5	3.10
6	3.45
7	3.81
8	4.17
9	4.52
10	4.88
12	5.59
14	6.30
16	7.01
18	7.72
20	8.43
24	9.86
28	11.28
32	12.70

without the necessity for regular repainting. Here is a summary of the available finishes:

- *Stainless steel* The best choice for hostile environments.
- *Fluoroplastic (PTFE)* A relatively new addition to the range of 'overcoatings', an alternative to zinc or cadmium plating but with better resistance to corrosion. Fluoroplastic coatings resist bimetallic corrosion where ferrous fixings are used to join other materials.
- *Sherardized or bright zinc plate* Both will survive 10 or more years without painting.
- *Electro-brassed* A shiny yellow copy of brass, used internally for fixing reproduction brass ironmongery.
- *Blued* Blued steel has a dark oxide coating with a temporarily protective lubricating oil finish. Not suitable for external use and has to be painted internally.
- *Japanned or Berlin* A cheap black enamel finish only suitable for internal use. Commonly associated with 'Tudorbethan' taste. Today there is a large range of decorative finishes to steel fixings, bronze coloured and plated (nickel, cadmium). Cadmium is toxic and should not be used in food preparation areas. Chromium-plated screws are commonly used in decorative work, but they need frequent cleaning to stay bright.
- *Bright steel* Rusts if moisture is present.

Screws can be classified by function and by head type. Refer to Fig. 5.17 for a description of the types of shank and screw head. The major development in recent years has been the introduction of self-drive and self-tapping screws with special heads for use with power tools. Also available are case-hardened wire grade or stainless screws that perform as if made from high-tensile steel.

General points to note on screw heads are as follows:

- *Countersunk* The heads are shaped to fit 'flush' in countersinkings in wood, or in metal components such as butt hinges.
- *Countersunk and raised* The raised heads reduce the danger of damaging surrounding surfaces in driving the screws.
- *Round* Particularly suitable for fixing metals which are too thin to countersink
- *Mirror* For fixing glass and other panels, the slots being concealed by screw-on or snap-on domes.
- *Square* For heavy-duty *coach screws*, in carpentry usually 6 mm diameter or larger, driven with a spanner.

Driving profiles are:

- *Slot* The standard type.
- *Clutch* The profile prevents removal.
- *Recessed* The *Phillips* and *Pozidrive* heads allow greater purchase to be applied in driving and the specially designed screwdrivers are less likely to slip. Recessed head screws are particularly suitable for mechanical driving.

Thread patterns are:

- *Single spiral* These traditional screws have two-thirds of their length tapered and threaded
- *Double spiral* Twin parallel threads on cylindrical shanks have greater holding power, and they extend over a greater proportion of the length of the screw.

For access panels, glazing beads and similar removable items round or raised head, slotted or recessed heads are appropriate. Brass or other metal, or plastic cups allow screws to be withdrawn and redriven without damaging the surrounding wood (see Fig. 5.17).

Where screw heads are to be concealed, they may be recessed below the surface and the hole filled with a proprietary filler. If the surface is to be clear finished, a *pellet* of the same timber with the direction of the grain following that of the surrounding timber may be glued in, providing a virtually 'secret', but permanent, fixing.

Turn-button See Figs 5.34 and 5.36 for turn-buttons of wood or metal. These are used to fix tops to tables and benches while allowing moisture movement to take place in one direction without attendant distortion or cracking of the tops or frames. In addition, tops can be removed if repolishing or renewal is required.

Slotted angle Slotted angle connectors allow movement

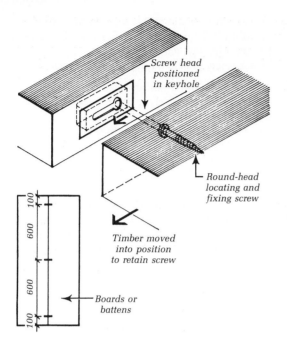

Figure 5.18 Keyhold joint with slot screw.

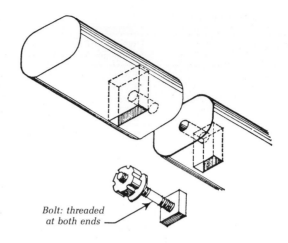

Figure 5.19 Handrail bolt.

in one direction but tops cannot be removed so easily. See Fig. 5.36 for details.

Slot screw   Figure 5.18 shows a slot screw, also used for joining board edges to form table tops, and for fixing panelling to framing and hanging cupboards on walls. A projecting screw on one member is inserted in the circular hole of a 'keyhole' cut in a plate which is fixed to a second member. The members are slid so the screw passes along the slot in the 'keyhole'. The plate then holds the screw head securely, and 'secretly', although, as glue is not used, the joint is demountable.

Bolts   With suitable washers these are used for heavy-duty work.

Handrail bolt   Figure 5.19 explains how a handrail bolt secures the ends of timber to make up a continuous handrail. A bolt threaded at both ends when tightened produces a very strong joint and which is difficult to detect in use. The slots on the underside of the timber are filled with matching timber and cleaned off smooth.

5.13 Demountable connectors

These enable preformed members to be rapidly located and joined. In addition to angle plates which allow moisture movement (see Fig. 5.32) and 'keyhole plates' which allow

dismantling (Fig. 5.18) other forms of connectors include the following types:

- *Corner plates*   with slots shaped so they draw parts together.
- *Two-part components*   which clip together, but which can be dismantled, others which draw parts together by cam action, and plates fixed to table tops with wood screws, threaded for metal screws in the tops of table legs.

Figure 5.20 shows a two-part plastics connector which holds members, usually boards, rigidly at right angles without the need for a housing or other framed joint, or for glue. If required, these joints are easily 'knocked down', although in fact their chief use is for rapid assembly of flat-packed furniture and kitchen cabinets. Using a template, the two members to be joined are each marked with two positions for screws; (a) shows the smaller part of the connector, with a nut placed in a recess in its upper surface, screwed to a horizontal board, (b) shows the larger part of the connector screwed to a vertical member, locked over the smaller part already fixed on the horizontal member, and then secured to it by a bolt.

5.14 Adhesives

(Reference: *MBS*: *Materials* Chapter 13.)

General guidance on the selection of adhesives for wood is provided in BRE *Digest* 340 (which supersedes 175 and 209). Guidance on gluing wood successfully is given in BRE *Digest* 314; and on gluing wood containing preservatives, in BRE *Technical Note* 31.

BS 1203 : 1979 covers the selection of adhesives for plywood and classifies them into four types, depending on

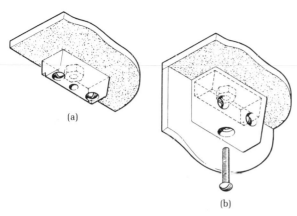

Figure 5.20 Patent connector. (By permission of Plasplugs Ltd, Burton-on-Trent, Staffordshire. Patent no. 1912/73.)

Table 5.9 Adhesives for exposures (Information from BRE *Digest* 175)

Category	Examples of exposure	Recommended adhesives	Durability of adhesive
Exterior			
High hazard	Full exposure to weather	RF, RF/PR, PF (cold setting)	WBP
Low hazard	Inside roofs or porches	RF, RF/PF PF, MF/UF	WBP BR
Interior			
High hazard	Laundries	RF, RF/PF, PF MF/UF	WBP BR
Low hazard	Inside dwelling houses, heated buildings, halls and churches	RF, RF/PF, PF MF/UF, UF Casein	WBP BR
Chemically polluted atmospheres	Swimming baths	RF, RF/PF, PF	WBP

MF = melamine formaldehyde; PF = phenol formaldehyde; RF = resorcinol formaldehyde; UF = urea formaldehyde; WBP = weather and boil-proof; BR = boil-resistant; MR = moisture resistant and moderately weather resistant; INT = interior.

their suitability for different types of application. These types are:

- WPB (weather- and boilproof): highly resistant to weather, micro-organisms, cold and boiling water and dry heat.
- BR (boil resistant): good resistance to weather, but will fail under prolonged exposure. Prolonged resistance to cold water and attack by micro-organisms.
- MR (moisture resistance): moderate resistance to weather. Will resist hot water for short periods and cold water for relatively long periods. Resistant to attack by micro-organisms.
- INT (interior): some resistance to cold water, but not necessarily to attack by micro-organisms.

BS 1204 uses the same classification of adhesives for wood. Part 1 deals with gap-filling adhesive types (0.15–1.3 mm), and Part 2 with close fitting joints (0.15 mm). In all cases, minimum performance and test data are specified. BS 4071 : 1966 (1988) covers PVAC (polyvinyl acetate) adhesives for interior use with wood; these are also covered in BRE *Digest* 340. There are three main kinds of glue used in the UK for joinery production. It is important to follow the prescribed working methods both off and on site, and all forms need clean dry conditions. The three subdivisions are as follows:

- *Animal glues* Known as Scotch glue and made from bones and hides. Useful for interior work, does not stain wood nor injure cutting blades of machinery. Not advisable in damp conditions.
- *Casein glues* Made from soured milk, available in powdered form for mixing with water. Stains timber, and difficulties arise in damp situations.
- *Synthetic resin glues* The most commonly used due to resistance to damp and to attacks from moulds and bacteria. There are three sources: urea formaldehyde, which has medium resistance to damp, and phenol formaldehyde and resorcinol formaldehyde, which are superior and reserved for external joinery and external plywood manufacture (details are given in Table 5.9).

The need to prepare surfaces so they are dry and clean and matched to give a thin *glue line*, for controlled curing conditions and with the exception of light duty contact adhesives, the need to clamp members together until the adhesive has set, limit the use of adhesives on the building site to work such as applying plastics laminates to existing surfaces. In workshops PVA and casein adhesives which are clean and easy to use have supplanted animal glues. In factories one- or two-part synthetic resin adhesives with the facilities and control available enable high-strength durable joints to be formed very rapidly, especially where radio frequency curing is used.

Surfaces to receive glue must be free from oil, grease or dust, and the glue film must be evenly applied. The manufacturer's particular recommendations should be followed. The moisture content of the timber and the

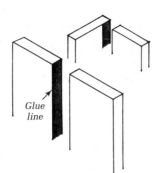

Figure 5.21 Butt joints.

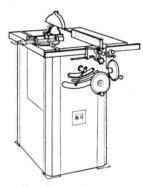

Circular saw bench

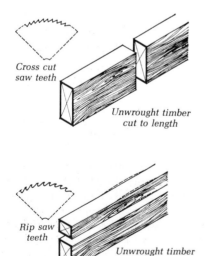

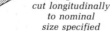

Figure 5.22 Sawing.

temperature of the room and of the glue are of paramount importance. To attempt to joint wet timber in a cold damp workshop or site shed is to invite failure.

Correct application and mixing of the catalyst glues are also vital, as is the observation of the correct procedure in bringing the pieces into perfect contact and cramping up. Fig. 5.21 shows simple *butt* or *rubbed* joints. Table 5.9 shows adhesives which are suitable for five exposure categories.

The description 'weather- and boilproof' (WBP) applied to plywood refers only to the bonding agent and although the adhesive is suitable for hazardous exposures, for equivalent durability the timber must either be heartwood of a durable species or be treated with a compatible preservative.

5.15 Joiners' shop production

In addition to mass-production joinery factories, joinery workshops are run by specialist firms and by contractors. Specialist firms and the large contractors maintain substantial timber stocks, and in some cases drying kilns. Some firms specialize in very high quality work for prestige buildings. In such workshops timber is thicknessed, planed, moulded, sanded and the joints are cut by machine. The parts will then be fitted together, wedged, glued, cramped and finished by hand.

A typical sequence of operations in the manufacture of timber components in a well-equipped joiner's shop will be as follows: the shop foreman 'takes off' the quantities of timber required for a component, from the architect's large-scale working details or from the workshop setting-out rod, and from these quantities prepares a cutting list. The information contained in the cutting list for each project is used when choosing the timber required from the storage racks. The amount of timber used will be set down on a *cost/value sheet*.

5.15.1 Machining

The basic processes are illustrated in Figs 5.22–5.29. The timber is first cut to length by means of a *circular cross-cut saw*. Then a *circular rip saw* is used to cut the timber to width and thickness. The timber is cut 'oversize', allowance being made at this stage for planing the finished timber, say, 3 mm on each face. The *standard method of measurement* allows a maximum of 3 mm. The sawn timber then passes to the *surface planer* to provide a true *face* and *edge*, and to remove any twist or irregularity. The timber then passes to a *thicknesser* which reduces the timber to the right size. The *surface planer* and a *thicknesser* are often combined as one machine as illustrated in Fig. 5.23.

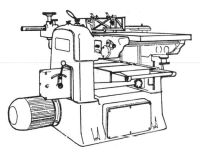

Figure 5.23 Combined planer and thicknesser.

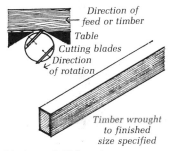

Figure 5.24 Planing and thicknessing.

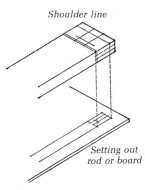

Figure 5.25 Setting out.

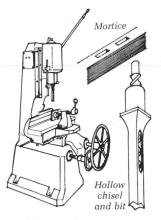

Figure 5.26 Morticing machine.

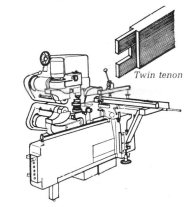

Figure 5.27 Tenoning machine.

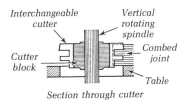

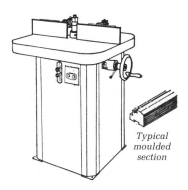

Figure 5.28 Spindle moulding machine.

At this stage the timber goes either to the site for fabrication, or to the setting-out bench for further machining. The setting out is done by the shop foreman, who draws the vertical and horizontal sections of the components full size on a plywood sheet or setting-out board, say 225 mm wide × 12 mm thick. This board is also known as a *setting-out rod*.

As an alternative to drawing direct on to plywood, full-size drawings can be prepared in negative form on transparent material such as tracing paper. This has the

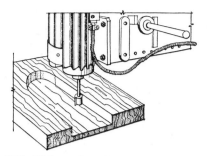

Figure 5.29 High-speed overhead router for cutting square or curved housings.

advantage that prints can be taken to provide a record of the work.

The setting-out rods, being full-size working drawings conventionalized by the shop foreman, go to the marker out, who transfers the relevant lines on to the timber. The dimensions from the setting-out rod can also be used by the shop foreman to produce the cutting list.

Lines of joints, and cuts with depths and type, are marked on the timber from the setting-out rod. Since proportions and types of joints should be to accepted standards, it is not always necessary to specify setting-out details of joints on working drawings. However, it would be wiser to include such details in the 'claims conscious' world of the larger contractor.

From the *setting-out bench* (Fig. 5.25) the timber travels to the *morticer* (Fig. 5.26) or the *tenoner* (Fig. 5.27), then to the *spindle moulder* (Fig. 5.28). The timber can be rebated, moulded, grooved or chamfered on the spindle moulder. Standard profiles are available but cutters can be specially made to any reasonable profiles, and circular work can be carried out.

Surface planer, thicknesser and spindle moulding operations can be combined on a *planer and moulder*. If this machine is used, the mortice and tenon operations are carried out later. The following machines are ancillary to the main production line machinery:

- *Router* for recessing to any profile as shown in Fig. 5.29.
- *Boring machine* for drilling holes in series.
- *Dovetail machine* for fabrication of dovetail joints, mainly used for drawer construction. It cuts front and sides together.
- *Panel saw* (or dimension saw) for cutting sheet materials, such as hardboard, plywood, etc. to size.
- *Band saw* for cutting circular work.

5.15.2 Sanding machines

- *Belt sander* which consists of a belt of abrasive sheet under which the timber is passed.

- *Drum sander* which consists of a number of drums covered with abrasive sheet over (or under) which the timber passes, on a moving bed.

After machining, the timber passes forward for fabrication. Portable power tools, such as a planer, drill and hand sanding machines — disc or orbital — will be used during this part of the work. See Fig. 5.30(a)−(e).

5.16 Factory production

Large firms mass-produce windows, doors and door sets, cupboard fittings, staircases and section profiles such as skirtings and architraves. Machinery is similar to that used in joiners' shops, but handwork is almost eliminated and production is large and rapid. There is likely to be an increasing demand for fully finished components, glazed, painted and complete with locks, hinges and fastenings. Refer to manufacturers' catalogues for computerized working schedules for door or window production in a factory.

The main processes of production are as follows. Timber, bought up to 12 months in advance, requires a large dry storage area. Scantlings are sorted automatically into equal lengths, and after selection for quality into 'sets' of say 160 pieces of 100 × 50 mm. These are stacked under cover for further air drying and then kiln drying takes from 2 to 18 days according to the species and sizes, with careful control of temperature and humidity throughout. Sawn sections of the appropriate sizes are cut to lengths to make the most economic use of the scantlings available.

Figure 5.31(a)−(c) shows the processes for machining the sills and jambs of windows (window manufacture has been chosen due to the relative complexity of components as compared with simple door joinery which is discussed in Chapter 6):

(a) The cut lengths passed on a conveyor, are planed to size and the initial cuts are made to form the required profile. Sill and jamb sections then go on separate lines.
(b) Mortice slots are cut in sills, and tenons are formed on the ends of jambs.
(c) A moulder completes the cutting of the profile, with grooves, throats and chamfers, as required. Sections are batched so as to minimize the changing of cutters. The maintenance and repair of woodworking machinery and the sharpening of cutters are done, wherever possible, automatically in a separate workshop.

Random lengths of moulded sections can be cut to non-standard lengths to form 'special' windows, which being out of the main production line are more expensive, although not always prohibitively so.

Sections for external joinery may be vacuum impregnated

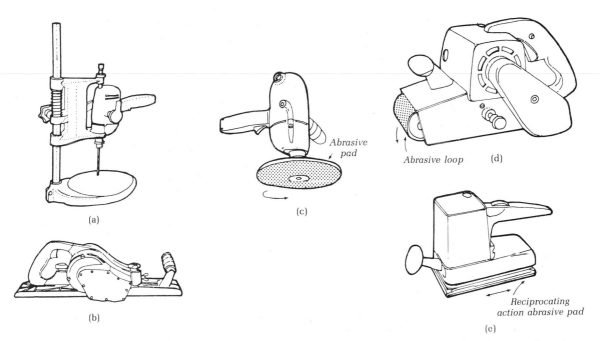

Figure 5.30 Portable electric power tools: (a) drill and stand; (b) planer; (c) disc sander; (d) orbital sander; (e) reciprocal sander.

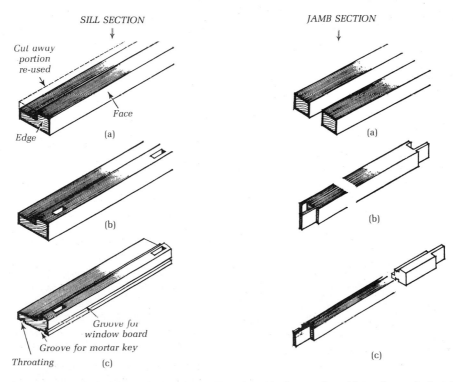

Figure 5.31 Machine fabrication of sill and jamb sections. Sill section: (a) planer and moulder to face and edge; (b) morticer cuts mortice slots as required; (c) moulder completes sill profile. Jamb section: (a) scantling, cut into section and passed through planer and moulder face and edge; (b) double end tenon cutter shapes tenons at each end of section; (c) moulder completes jamb profile.

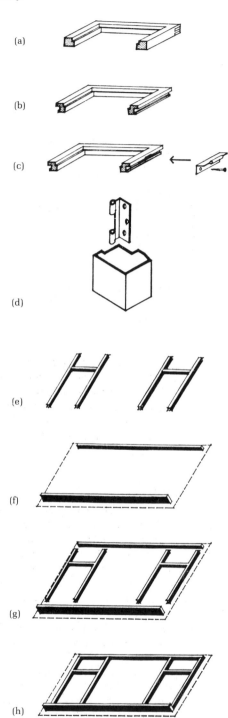

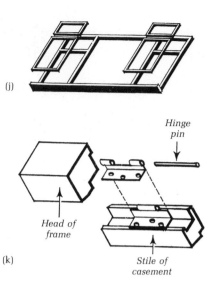

Figure 5.32 Manufacture of standard window.

with preservative before assembly. See *MBS: Materials* Chapter 2. In addition to random checks, parts are inspected at several stages:

- Length sorting
- Cross-cutting
- Cleaning off
- Finishing moulder

Figure 5.32(a)–(k) shows stages in the manufacture of the standard window shown in Fig. 5.33(a). Designers will complain that the combination of the top-hung and side-hung sash is one of the ugliest reminders of British Standard windows. The complexity of the top- and side-hung forms is, however, a useful demonstration of fabrication methods. The frames, sashes or casements and vents are fabricated separately.

The stages in the manufacture of the standard window shown in Fig. 5.32(a)–(k) are as follows:

(a) The timber sections are machined to form rebates to receive the glass, and then cut to length. The ends of the top rails, stiles and bottom rails of the casement are then machined to form combed joints. The loosely assembled casement is placed in a jig with a cramping device which squares and lines up the joints, glues them with synthetic resin glue and automatically drives a pin or star down at the corners.

(b) The assembled casement is passed through a drum sander and then through a moulder which profiles the outer edges.

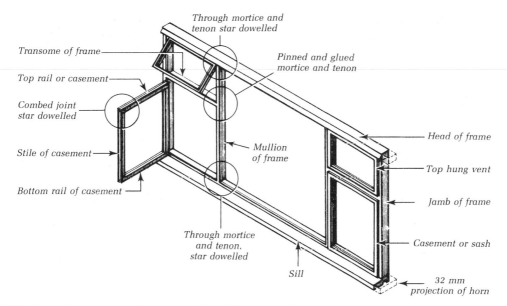

Figure 5.33 Standard casement windows as manufactured by process illustrated in Fig. 5.32.

(c)   Recesses are cut for cadmium-plated hinges. The hinge consists of two interlocking cranked portions, and a separate pin. The single knuckle part of the hinge is dropped in place in the recess and machine screwed into position by zinc-plated screws. The casement is dipped in preservative. The vacuum process cannot, in this instance, be used since the face profile is cut after assembly.

(d)   The double knuckle part of the hinge is screwed into position on the jambs of the frame.

(e)   The transoms are fitted into the jambs to form H-frames, which are then glued.

(f)   The head and sill are placed in position on an automatic cramp bed.

(g)   The H-frame is set down, on the cramp bed, the joints between the H-frame and the head and sill being open and spaced slightly apart.

(h)   The tenons are glued from a hand dispenser and the frame is cramped up.

(i)   The component is passed through a drum sander.

(j)   Pre-assembled casements are placed in position on the frame.

(k)   Hinge pins are driven home by a vibrating hammer.

(l)   (Not illustrated.) Knots are sealed (knotted) and the casements wedged open by metal dogs. The whole frame is immersed in primer and passed through a dryer. Casements are temporarily secured by plywood battens and the windows are stacked before delivery.

The other window forms illustrated in Fig. 5.33(b) and (c) demonstrate high-performance sashes complete with ironmongery in place. The single length of timber needed for head and sill members is the limitation upon the width of frame that can be assembled. Wider units will be formed by coupling frames using cover battens to jambs and cover flashings to weather sill junctions.

5.17 Joinery fittings

Fittings such as bar fittings, work benches and cupboards can be 'box framed' from suitably thick boards which provide general solidity and make for easily framed joints. Alternatively, the frames are made with slender members (either front frames or cross-frames as Figs 5.34 and 5.35) which are infilled or clad with thin sheets of ply or board. The box frame method is economical for flat-pack cabinet work (kitchens and wardrobe units) and for one-off production. The second method will decrease the weight of joinery fittings and was the traditional mass production technique before ply, chipboard and particularly MDF took over the role of framing in cabinet work. Figures 5.34 and 5.35 are included to show with three-dimensional drawings the implications of three-way intersections between members in traditional framing. Fittings should be designed so as much fabrication as possible can be carried out in the joiner's shop in order that site labour and time in fitting are minimized.

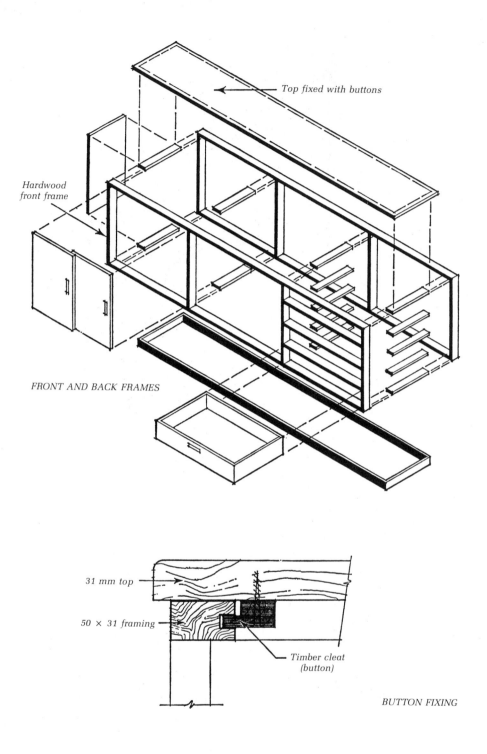

Top fixed with buttons

Hardwood
front frame

FRONT AND BACK FRAMES

31 mm top

50 × 31 framing

Timber cleat
(button)

BUTTON FIXING

Figure 5.34 Cupboard units with front and back frames.

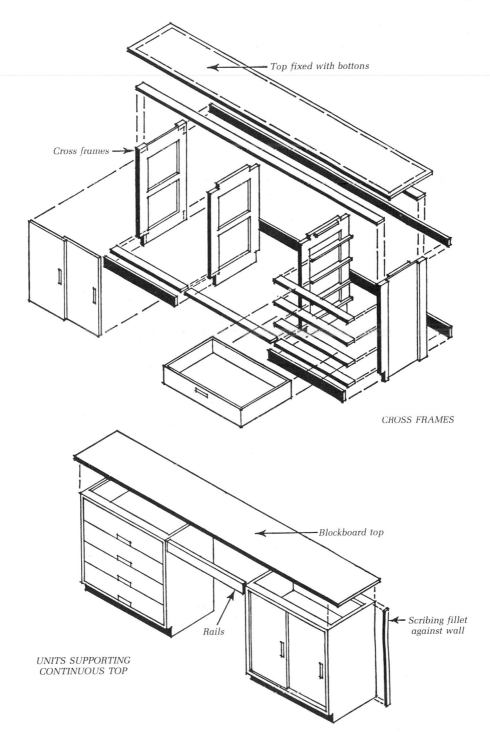

Top fixed with bottons

Cross frames

CROSS FRAMES

Blockboard top

Scribing fillet against wall

Rails

UNITS SUPPORTING
CONTINUOUS TOP

Figure 5.35 Cupboard units with cross-frames and with units below top.

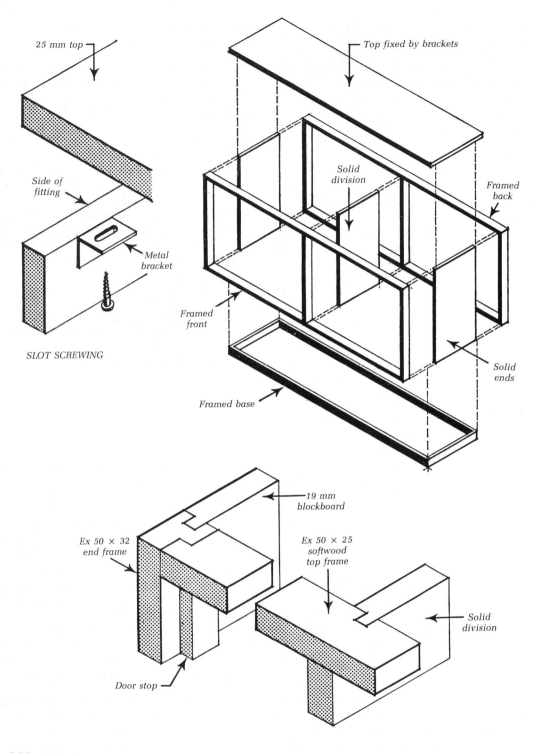

25 mm top

Top fixed by brackets

Side of
fitting

Metal
bracket

SLOT SCREWING

Solid
division

Framed
back

Framed
front

Solid
ends

Framed base

19 mm
blockboard

Ex 50 × 32
end frame

Ex 50 × 25
softwood
top frame

Solid
division

Door stop

Figure 5.36 Cupboard units using box-framing principles.

In public buildings special consideration must be given to wear on table tops and the effect of scuffing and floor cleaning operations on bases to fittings, for which timber is not always a suitable finish. Tops can be faced with laminates, the better wearing finishes having real timber veneers incorporated within a plastic laminate. Other well-tried facings include linoleum, vinyl sheet (as used for flooring), granite/marble and agglomerate materials like terrazzo or 'Corian' which can be cast to resemble marble. Brass-faced worktops are also feasible with metal dressed and glued over timber profiles. Stainless steel worktops and drainers are available to standard profiles up to 3 m length and are carried on blockboard. Plinths or skirtings should also be faced in the sheet materials already mentioned; recessing the skirting to form a 'toe space' is more satisfactory for ergonomic reasons.

5.17.1 Cupboard units

It is best to consider a traditional framed fitting as comprising vertical frames in one direction joined by members in the other direction. Thus, Fig. 5.34 shows front and back frames joined by cross-members. These frames stand on a base, part of which is infilled to form the bottom of a cupboard. A top of cross-tongued boards or veneered blockboard would be fixed, either with 'buttons' or with screws in slotted angles — methods of fixing which allow differential movement. Buttons also permit easy removal and replacement of the top. Ends and vertical divisions are formed by panels with stiles and rails either infilled with plywood or covered to give a 'flush' appearance. Figure 5.35 shows cross-section frames to receive longitudinal rails and drawer sides and separate units below a top.

A further method shown in Fig. 5.36 is to make the fitting on a 'box framing' principle with solid ends and panels, out of 19 or 25 mm nominal blockboard having edges lipped by machine or veneered, with a thin plywood back for bracing. A light framework is made for the front and back. The cupboard units to be fitted complete with all furniture in the joiner's shop. This pattern simplified to shells made from MDF or chipboard and joined by mechanical connectors and hung with doors using lay-on hinges is the basis of 'flat pack' cabinets favoured by kitchen cupboards and furniture makers today.

The remaining illustrations on modern joinery practice are taken from outstanding examples of detailing in timber from the work Evans and Shalev, Peter Schmitt and John Partridge of HKPA, Dennis Lennon & Partners, and the Robin Wade & Pat Read Partnership; full credits are given with the illustrations.

5.18 Case study: Truro Crown Courts

Architects: Evans and Shalev, 1988.

The Truro Courts of Justice (Fig. 5.37) were selected by *The Architects' Journal* as the 'Building of the year' in 1988, and the excerpts published have been based upon the *AJ* issue of 28 September 1988 by kind permission of Evans and Shalev.

The three court interiors are composed to similar elements, the selected details relating to the panelled ceilings and jury benches. Services were so arranged that demountable ceilings were not required, the panels comprising fire-resistant boards with applied softwood mouldings fixed over preservative-treated framing of 100 × 50 mm members. The ceiling decoration depended upon eggshell oil paint, the effectiveness of the joinery with regard to minimizing shrinkage relying on glued and pinned assembly, under warm dry conditions on site.

The purpose-made benches are constructed from MDF board veneered in ash, with solid ash profiles employed for mouldings and nosings; the finishing comprises a grey stain and lacquer. The supporting structure is a steel tubular frame, welded to T's and flats to provide a robust framework with minimal members.

5.19 Case study: Warrington Crown and County Courthouse

Architects: Howell, Killick, Partridge and Amis, 1992.

A high standard of internal finishes has been used throughout. All joinery was purpose made and constructed from mahogany obtained via managed forest estates in Brazil. The court rooms and related entries have fibrous plaster ceilings, both coffered and vaulted and integrated with the heating and air-conditioning system. Mahogany has been employed for benches, fitted furniture and panelling. The sample work chosen for illustrating are the robing lockers, and library shelving system designed on a modular approach (Fig. 5.38). HKPA practice has a long tradition in excellent joinery detailing and which dates back to their work for colleges at Oxford and Cambridge, for example St Anne's and St Antony's (Oxford) and Darwin, Downing and the University Centre (Cambridge).

Further examples of detailing which follow this logical and sequential pattern can be seen in the case studies dedicated to doors (Chapter 6) and stairs (Chapter 8).

5.20 Case study: Rothschild International Bank, Moorgate, London

Architects: Dennis Lennon & Partners.

The joinery details (Fig. 5.39) are taken from the boardroom design and represent the highest standards in finish and quality of materials. The bookcases and door trim are illustrated to reveal the way a designer develops a language with similar materials and their juxtaposition in cladding and framing the work. It is the use of stainless steel trim in the jambs and edgings of doors and as thin separators between panels which marks Lennon's work as that of a superior detailer. The use of shining metal in this way not only gives an impression of incontestable quality, but asserts the discontinuity between the components. The work achieves the distinction of Grey Wornum's joinery

(a)

Figure 5.37   Truro Crown Court. Ceiling panels and jury bench details: (a) court interior (photo: Martin Charles); (b) details of ceiling construction; (c) ceiling (photo: Martin Charles); (d) bench framing and cladding. (Courtesy *The Architects' Journal*.)

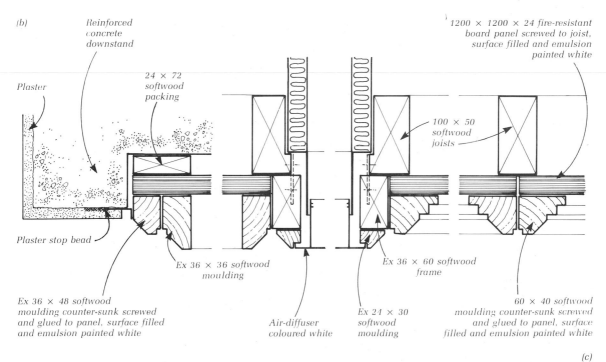

(b)

Reinforced concrete downstand

Plaster

24 × 72 softwood packing

1200 × 1200 × 24 fire-resistant board panel screwed to joist, surface filled and emulsion painted white

100 × 50 softwood joists

Plaster stop bead

Ex 36 × 36 softwood moulding

Ex 36 × 60 softwood frame

Ex 36 × 48 softwood moulding counter-sunk screwed and glued to panel, surface filled and emulsion painted white

Air-diffuser coloured white

Ex 24 × 30 softwood moulding

60 × 40 softwood moulding counter-sunk screwed and glued to panel, surface filled and emulsion painted white

(c)

at the RIBA Building where metals and veneers define surfaces but in a way demonstrating a finer and more delicate side. It should also be noted that the metal strips are used to isolate vulnerable materials such as cloth or leather, which may need replacement earlier than the rest of the installation.

5.21 Case study: The Jameson Irish Whiskey Centre, County Cork, Eire

Designers: Robin Wade & Pat Read Partnership.

The assembled shop counter details (Fig. 5.40) reveal the method of communicating the designer's information to the joiners with key elevations showing the arrangement and sizes with typical sections through the differing patterns of counter cupboard (drawers, panel front, glass front etc.). The sizing of members were intended for Parana Pine but during manufacture it was decided to use the tropical hardwood of a non-endangered species. The framing varies from traditional tenoned frames to MDF box-framing with facings to the public side to simulate solid framing. The jointing and finishing notes are not given since these are part of the sub-contract specification.

Notes

1   *Self Build and the Art of the Carpenter*, Green Books, 1991. 'The Self-build Book' by Jon Broome and Brian Richardson is an

Figure 5.37 *continued*

excellent résumé of the skill exercised by self-builders and develops a number of interesting details of carpentry framing and joinery work.

House by Tracy Kidder, Picador, 1986, describes the process of framing and cladding a timber house from the point of view of the workforce.

2   Refer to joinery details using limited profiles in John Sergeant's book *Frank Lloyd Wright's Usonian Houses*, New York, 1976.

3   TRADA publication, *Mechanical Fasteners for Structural Timber Work*, (1985). *AJ* feature, 'Fixings and adhesives', supplement for July 1983.

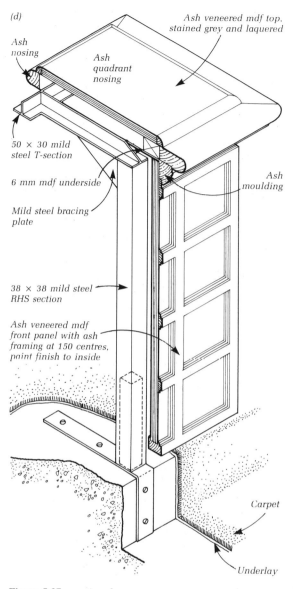

(d)

Ash veneered mdf top,
stained grey and laquered

Ash
nosing

Ash
quadrant
nosing

50 × 30 mild
steel T-section

6 mm mdf underside

Mild steel bracing
plate

38 × 38 mild steel
RHS section

Ash veneered mdf
front panel with ash
framing at 150 centres,
paint finish to inside

Ash
moulding

Carpet

Underlay

Figure 5.37 *continued*

(a)

(c)

Figure 5.38 Warrington Crown and County Courthouse.
Judges' library: (a) view of installation (photo: HKPA); (b)
part elevation and inset isometric; robing lockers: (c) view of
locker group (photo: HKPA); (d) section and elevation;
courtroom bench and panelling: (e) view of judges' bench.

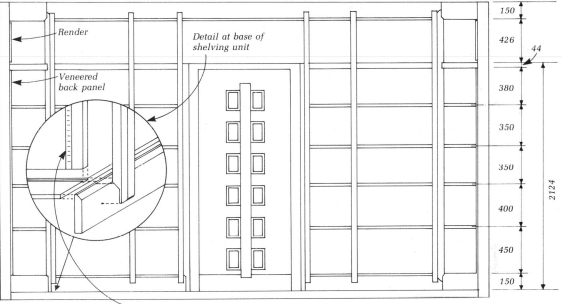

Render

Detail at base of
shelving unit

Veneered
back panel

150

426

44

380

350

350

400

450

150

2124

(b)

Recessed tonk strips or equivalent — adjustable
shelving let into vertical panel of veneered ply

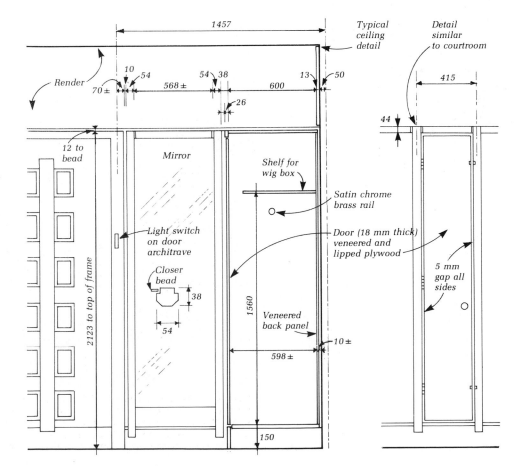

1457

Typical
ceiling
detail

Detail
similar
to courtroom

Render

10
54

54 38

13 50

70±

568±

600

26

415

44

12 to
bead

Mirror

Shelf for
wig box

Satin chrome
brass rail

Light switch
on door
architrave

Door (18 mm thick)
veneered and
lipped plywood

Closer
bead

38

5 mm
gap all
sides

54

2123 to top of frame

1560

Veneered
back panel

10±

598±

150

(d)

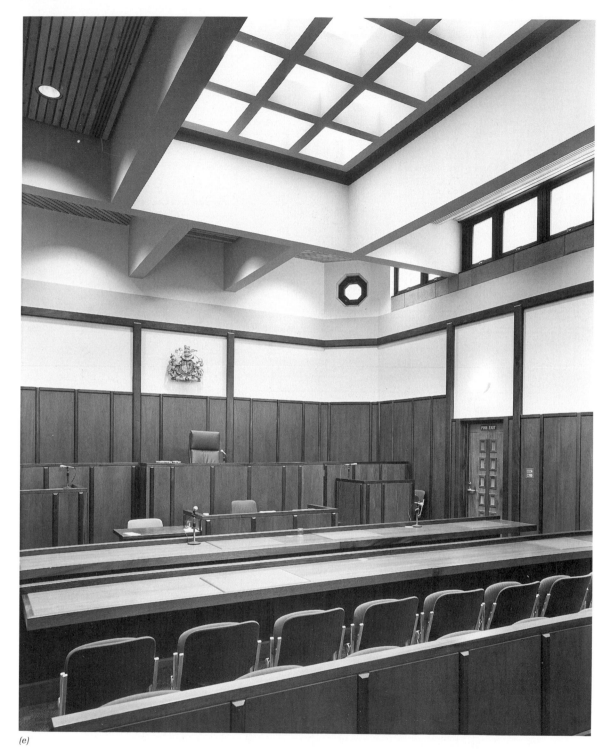

(e)

Figure 5.38 *continued*

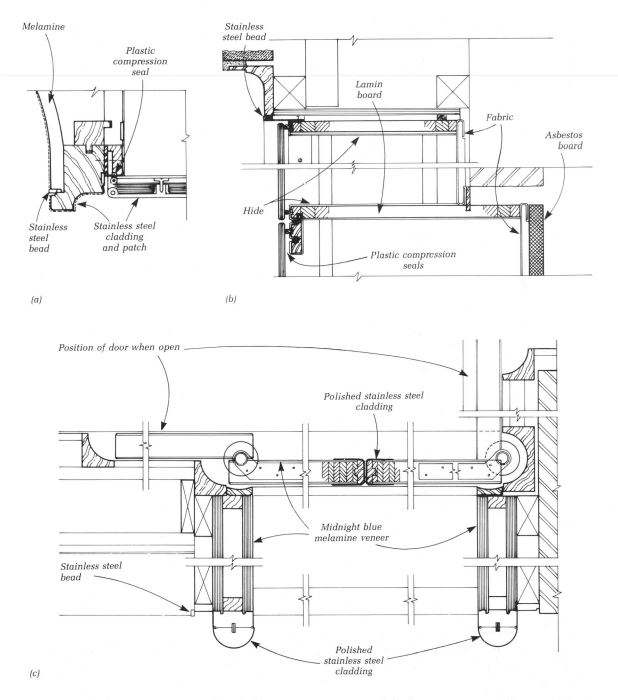

Figure 5.39 Rothschild International Bank. Joinery details: (a) plan at corner of bookcase; (b) section through bookcase; (c) plan of door to boardroom. (By kind permission of *Architectural Review*)

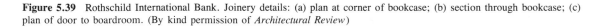

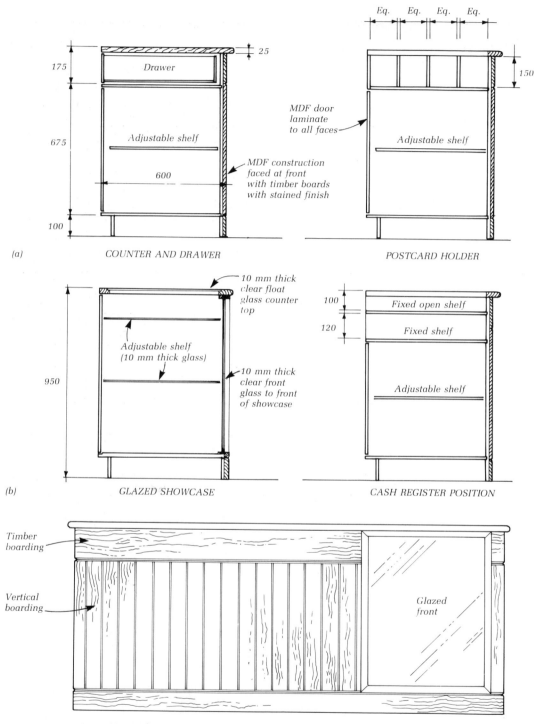

Figure 5.40 The Jameson Irish Whiskey Centre. Shop counter details: (a) typical framing sections; (b) typical front elevation; (c) alternative front elevation; (d) typical elevations: end and cross-section; (e) typical plan. (By kind permission of the designers.)

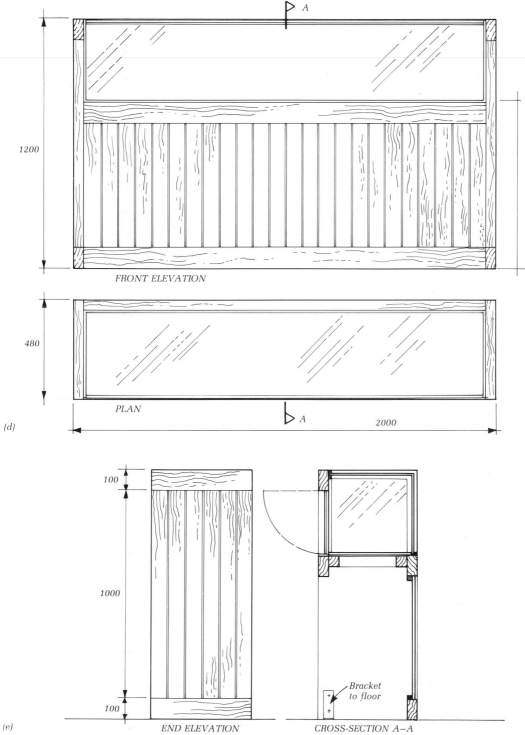

FRONT ELEVATION

1200

A

PLAN

480

2000

(d)

100

1000

100

END ELEVATION

CROSS-SECTION A–A

Bracket
to floor

(e)

6 Doors

6.1 Introduction

Doors are one of the most heavily used and abused movable components that occur in a building. They share the distinction of being on constant duty alongside flushing cisterns, bibcocks and ball valves, apart from mechanical installations such as fans and lifts, etc. Doors have to be carefully detailed and specified to respond to usage. It is important for the designer to gauge the way building users will perform — looking at their previous abode will help. Making a wrong assessment could be compared with Swedish door sets first installed at the Horse Guards Barracks, Knightsbridge, in the 1960s. The clear pine doors lasted 12 months due to the guardsmen's habit of returning off duty carrying two handfuls of equipment (busby, coat and sabre), the doors being eased open by a sharp clout from a spurred riding boot! The previous establishment had solid framed doors which had lasted a 100 years or more. By comparison, conditions imposed by temperature and humidity are relatively easy to handle.

Internal doors are either made to traditional framing patterns or more commonly of flush construction. External doors are illustrated in *MBS: External Components* Chapter 5 and include yard gates and the more robust and secure forms of entry and service doors needed at building entrances.

Doors within buildings often perform significant roles in preventing the spread of fire, and such elements have to be selected in relation to the enclosure of frame or screen. A common situation is the inclusion of fire-resisting doors within the compartment walls to the interior. Doors are capable of fire-resistance standards from 30 minutes to 4 hours.

6.2 Performance requirements

Returning to the concept of doors as 'movable components' it is worth considering the forms available and illustrated in Fig. 6.1. The method of hanging doors can be defined as pivoted, hinge carried or track supported. For the large openings, overhead track support and roller shutters provide trouble-free running and can be power operated. Sill- or floor-based working tracks should be avoided as they are prone to damage and need high standards of upkeep. Revolving doors provide draught-free entries and are in many ways superior to a lobby approach. Security versions are also made which only permit accredited access, sometimes used to provide high security within buildings. The whole topic of revolving doors is covered by *MBS: External Components*.

British Standards are in urgent need of updating. The following standards are useful current references: BS 449 (revised 1988) Part 4 only refers to match-boarded doors.

BS DD 171 : 1987 *Guide to specifying performance requirements for hinged and pivoted doors.*
BS 4787 : 1980 *Internal and external wood doorsets, door leaves and frames.* Replaces Parts 1, 2 and 3 of the former BS 449.
BS 6262 : 1982 *Covers safety glass in glazed doors and screens.*
BS 1245 : 1986 *Metal door frames (steel).*
BS 5286 : 1978 *Aluminium framed sliding doors.*
Code of Practice 151 : Part 1 : 1957 describes 'Wooden Doors' but this is largely craft based and out of date except for fixing details.

The PSA has produced a useful guide in their series 'Method of Building' titled *Performance Specification Internal Doorsets*. This should be read in conjunction with BS DD 171 : 1987. Refer to 'Fire doors' in section 6.9.1 for further notes on British Standards.

6.2.1 Appearance

Doors and doorways play a significant role with the first

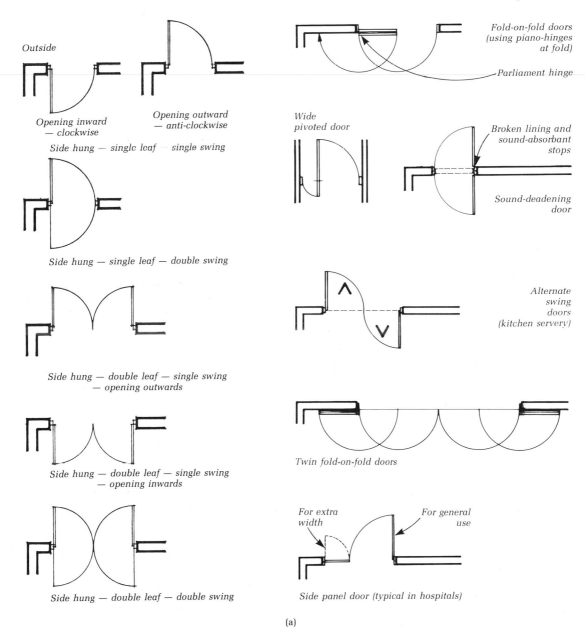

Figure 6.1 Methods of hanging doors: (a) hinged; (b) sliding; (c) folding; (d) revolving.

impressions gained in entering internal spaces. Questions of solidity or transparency arise as well as matters of proportion, scale and tactile quality. A compendium of these features can be found in old encyclopaedic volumes like *Doors and Gates* (by Albert Haberer, published through Illife Books Ltd in 1964).[1] These will still be more inspirational than many suppliers' catalogues. Haberer reveals the range of construction that can be expressed in

the detailing of doors and doorways together with architraves and framing that can form a significant part of the design.

6.2.2 Durability

The resistance to 'wear and tear' and strength are the critical factor for internal doors, although joinery in damp situations

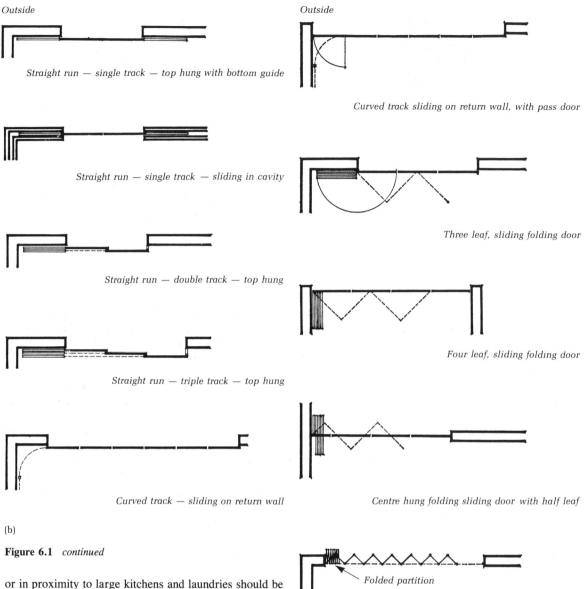

Outside

Straight run — single track — top hung with bottom guide

Straight run — single track — sliding in cavity

Straight run — double track — top hung

Straight run — triple track — top hung

Curved track — sliding on return wall

(b)

Outside

Curved track sliding on return wall, with pass door

Three leaf, sliding folding door

Four leaf, sliding folding door

Centre hung folding sliding door with half leaf

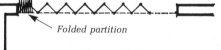

Folded partition

Collapsible, on top track and bottom guide as for a metal folding shutter gate

Flexible, on top track — use as an internal partition or space divider

(c)

Figure 6.1   *continued*

or in proximity to large kitchens and laundries should be clad with impervious facings with frames or linings terminating above impervious skirtings. Stainless and coated steel doors are made for industrial applications and form typical assemblies for lift and landing gates. For other aspects of durability and weatherproofing refer to 'external doors' in *MBS: External Components* Chapter 5.

6.2.3 Definitions

Door leaf Movable part of door set.
Door frame Fixed framework, built into or fixed to structure of building. Designed for receiving and hanging door.

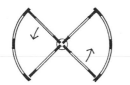

Revolving position for draught exclusion

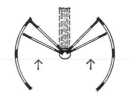

Leaves collapsing against pressure from crowd

Leaves folded flat to give clear passage

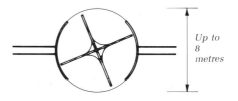

Up to 8 metres

(d)

Figure 6.1 *continued*

Door set Complete component of door leaf and frame, fanlights, etc. often supplied complete with fitted glazing and ironmongery.

Leaf frame Those parts of the door leaf which require mechanical strength.

Core A simple or composite material placed between leaf facings of door.

Facings External surface of door leaf enclosing core and leaf frame.

Lippings Finishing strips applied to edges of door leaf, usually to stiles but to four edges for high-quality veneered doors.

6.2.4 Doors to control air movement, or to limit passage of sound or thermal effect

Air movement A combined stop and compression seal is shown in detail in Fig. 6.2(a). This type of seal is used where an airtight or dustproof seal is required as in a computer room. The door must have a threshold and be sealed all round and top and bottom bolts should be fitted to hold the doors close against the seal.

Sound movement Details in Fig 6.2(b) and (c) show the way rebated door profiles and enlarged door stops can achieve superior sealing at door edges. The term for such doors should be 'sound resisting' — extra density (namely weight) in the construction will assist in controlling the passage of sound. Two sets of doors separated by a lobby or air gap will be needed in critical situations and where discontinuity is needed in the frame and wall construction. BS 8233 provides recommended figures for decibel reduction of doors according to the insulating value of the adjacent partition. Table 6.1 provides the relevant data for decibel reductions between 15 and 40 dB. Refer to section 6.14 for further references and details.

Thermal insulation The loss of heat through doors can be high although the provision of weather-seals in relation to entry lobbies will improve the situation. The only effective solution with doors in frequent use is to devise a lobby approach where heat conservation is required, the air locking meaning that minimal loss occurs. Revolving doors will also give effective control with heat loss. There are both metal- and ply-faced flush doors made with cavity-filled insulation which can provide excellent thermal characteristics.

6.2.5 Fire doors

Fire doors have a key role in fire precautions, the categories falling into three divisions:

1. *Openings within compartment walls* Openings permitted within walls separating compartment zones have to be constructed in accordance with the standards of fire resistance laid down within the Building Regulations. Compartment walls occur at the edges of fire zones and as envelopes to stair shafts.

2. *Means of escape doors* Such doors overlap in function with those employed in compartment walls, but are specifically related to means of escape routes. The legislative requirements vary considerably according to the pattern of building use. The references are complex and need to be agreed with the local fire department.

Table 6.1   Insulation values of partitions with doors from BS 8233

Construction	Insulating value of partition (dB)					
	25	30	35	40	45	50
Any door with large gaps around edge (15 dB)	23	25	27	27	27	27
Light door with edge sealing treatment (20 dB)	24	28	30	32	32	32
Heavy door with edge sealing treatment (25 dB)	25	29	33	35	37	37
Double doors with sound lock (air space or lobby) (40 dB)	25	30	35	40	44–45	49

Based on area of door being 7 per cent of area of partition.

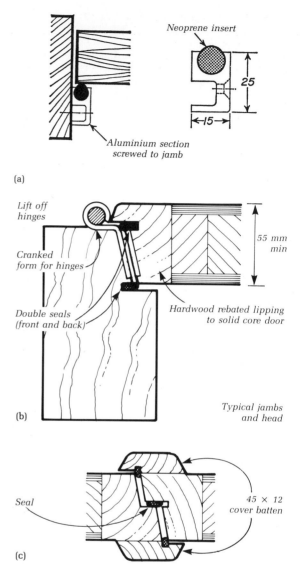

(a)

(b)

(c)

Figure 6.2   Improved door rebates and door stops: (a) timber door step; (b) double rebated door set; (c) treble rebate at meeting stiles.

The designer prepares floor layouts that can convey the basic information as to the agreed door sizes, fire door category, door furniture, emergency lighting and signing. Such layouts will need to be established formally with the fire department and to be updated and reconfirmed as sanctioned alterations occur.

3. *Smoke-stop doors and screens*   These provisions will be made to prevent the movement of smoke through a public building during a fire outbreak. They relate to escape routes and especially those involving stair shafts and basements.

4. *Other details*   Refer to details of fire doors and screens in section 6.9.

6.2.6 Strength and stability

Internal doors need to be robust enough to withstand the constant use inflicted. Normal domestic use will incur impact with furniture apart from scuffing at edges, hence the use of lippings to minimize damage at opening stiles. Doors in public buildings will suffer much heavier use and need to be constructed with stouter framing, say 44 mm minimum and with dense cores if constructed with ply facings. Lippings to four edges of flush doors will improve resistance to damage. Metal kicking plates and finger plates will combat wear and tear at the main points of contact. Full laminate facing or sheet metal cladding will provide a higher standard of protection and are commonly employed in industrial premises.

The most useful tests have been conducted by the BRE, PSA and DHSS; the findings are available in the British Standards document Draft for Development (DD) 171 : 1987. The document states four categories of door use, severe, heavy, medium and light duty, and sets down the impact energy to be resisted (Table 6.2). By comparison, the guardsman's boot (see section 6.1) can develop 4 N m while a shoulder charge can develop 100 N m. This also governs the selection of locking and

latching mechanisms. In broad terms severe and heavy-duty use need solid cores in flush doors, or else traditional framing with beaded flush panels (see Fig. 6.3). The greater size of members in traditional construction with frames and rails assuming 145 and 250 mm girth will lead to higher stresses due to moisture movement. Flush construction is easier to devise to achieve balanced movement stresses owing to the composite layers. Facings if needed to resemble panel designs are arranged as 'false cladding' glued to the ply skins. Such arrangements are often called for to fulfil fire rating of doors in historic buildings. This

Table 6.2 Door strengths from D 171 : 1987

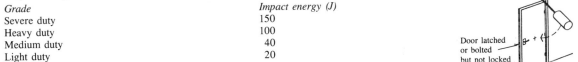

Heavy body impact
A sandbag of 30 kg mass is swung as a pendulum from varying heights to impact at centre of leaf width level with handle. The body is swung three times to both sides of leaf and the door must resist the specified force.

Grade	Impact energy (J)
Severe duty	150
Heavy duty	100
Medium duty	40
Light duty	20

Door latched or bolted but not locked

Hard body impact
A 50 mm steel ball is dropped onto the door face in a pattern to test all parts of the face. Heights are set to give specified impact energy.

Grade	Impact energy (J)
Severe duty	8
Heavy duty	5
Medium duty	3
Light duty	2

Release device

Failure is rupture of facing material not indentation

Table supports 3 edges only

Torsion
A force is applied normal to the plane of open leaf at mid-height on the lock side of the leaf. After preloading a force of 400 N is applied in increments of 100 N. The residual deflection of the lock side lower corner is measured after the force has been removed.

Grade	Deformation (mm)
Severe duty	3
Heavy duty	3
Medium duty	3
Light duty	3

Clamp to hold top of door rigid

Force applied

Deformation is measured here

Closing against an obstruction
A rigid block 50 × 50 × 10 mm is inserted between the door leaf and frame and a force of 200 N is applied as above in 50 N increments. There should be no damage to the door or its fittings which impair its action.

Grade	Applied force (N)
Severe duty	200
Heavy duty	200
Medium duty	200
Light duty	200

Force applied

Block inserted here

Note: Tests also specified for slamming shut and open, downward deformation and abusive forces to ironmongery.

can be achieved by slicing existing panelled doors to make a composite construction over a fire-resisting core. Traditional framing is illustrated in the joinery detailed by Howard, Killick, Partridge and Amis for the Hall of Justice (Fig. 6.29).

6.3 Framed and panelled construction

A door made with framing and panelling is the traditional form of construction. Successful design depends upon the proportions of panels. Recourse to historic reference or to 'copy books' will be needed to arrive at the appropriate design. Figure 6.3 provides a composite elevation to demonstrate the various forms of traditional panel. Accurate framing and well-made joints are essential if such joinery

is to perform in a satisfactory manner without warping or loosening of the panels. The method of construction is as follows:

* Stiles run top to bottom, with sufficient girth and thickness to ensure good fixings for morticed locks and for hinge security (say 85 × 44 mm for internal use and 100 × 44 mm externally)
* Rails framed into the stiles using mortice and tenon joints as Fig. 5.6. Top rails sized as stiles but bottom and lock rails of minimum 200 mm girth internally and 250 mm externally. Subsidiary rails of 95 mm or smaller dimension.
* The vertical middle rail or muntin or other vertical members are subtenoned into the horizontal rails. It is

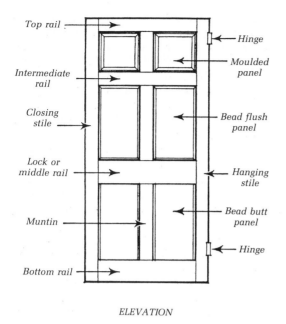

ELEVATION

Figure 6.3 Panelled door with composite details to show varying designs.

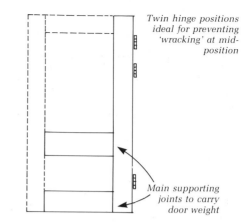

Figure 6.4 Key framing joints, twin-light or twin-panelled door.

customary to haunch the rails into the grooved stiles to prevent deformation of the horizontal rails (for definition of haunch refer to Fig. 6.6(b)). Doors are carried by the framing joints on the hinge side (head rail to stile). It would be logical to use the larger sizes for such members and to reduce the lock stile to the minimum for lock housing (Fig. 6.4). The pattern and size of lock are critical for the lock rail and stile junction and the tenon arrangement should be designed to accommodate the lock case. Alternative arrangements are of course feasible by mounting the lock in portions of the lock stile above or below rail joints.

6.4 Standard panelled and glazed doors

These mass-produced patterns follow the designs that conform to BS 4787 : Part 1 and illustrated in Fig. 6.5. Imported doors are made that comply with constructional standards, but offer greater variety of design. Security glass such as multi-ply has to be considered in glazed doors for public buildings as well as panels at vulnerable locations in domestic installations. In these days of litigation it is sensible to specify security glass for all internal glazing below 1.95 m level for all classes of work.

Turning to other components for panelled doors, these have to comply with the following requirements:

- Timber for framing and plywood for panels to conform

with BS 1186 : Part 2 : 1971). Adhesives comply with BS 5442 : Part 3 : 1979.
- Framing methods are optional; either dowelled joints or mortice and tenon, as follows. Dowels to be 16 mm hardwood, spaced at 57 mm centre to centre with three dowels at bottom and lock rail joints and two otherwise where mortices and tenons are used, then through haunched and wedged tenons at top, bottom and middle rails. Intermediate rails are sub-tenoned (min. 25 mm) into the stiles. Solid panels are formed in ply, framed tightly into grooves, with 2 mm spare width to allow for door shrinkage. For details of joints refer to Fig. 5.6 and for moulded members to Fig. 5.5.
- Finished sizes are given in British Standards specification, namely 85 mm girth for stiles and top rails with 175 mm girth for bottom and lock rails.
- Mouldings for glazing beads are delivered loose for pinned fixings on site.

In conclusion, the British Standard for cheap panelled and glazed doors, represents minimum requirements for low-cost building. Flush doors will generally provide less trouble in warping, the half glazed version available in flush doors also being less liable to accidental breakage than fully glazed panelled doors of either single- or double-panel design. Low-cost framed joinery is usually made for paint finish, with knotting and priming carried out at the joinery works. It is essential that quality inspections are made before priming occurs. Clear finishing is possible provided higher standards are called for in timber selection and workmanship. Old panelled doors which were made to thicknesses of 48–57 mm are far superior to British Standards which today permit leaves as thin as 40 or 44 mm. The other advantage with recycled joinery is the superior quality of softwood employed, often knot-free larch

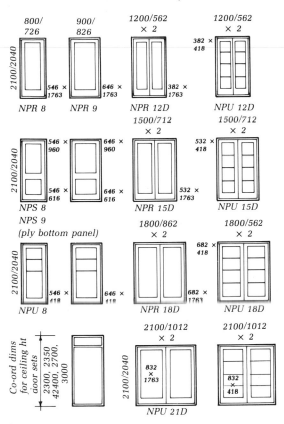

Dimensions: co-ord/leaf

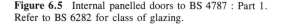

Dimensions of framing members are for all doors.
Glass dimensions (small type) are sight sizes.
Door leaf thicknesses: 40 × 44 mm.

Figure 6.5 Internal panelled doors to BS 4787 : Part 1.
Refer to BS 6282 for class of glazing.

or Douglas fir in nineteenth-century doors. The style of decoration in that century was either paint or 'grain and varnish'; clear finishing recycled joinery is often disappointing in appearance because of glue lines; as well as not being true to the decorative character of the period. The disadvantages in reusing old doors are the unusual sizes and the fact that fireproofing may be costly. It is also a matter of being selective, since in domestic work the best Victorian joinery was reserved for the ground and first floor with thin 40 mm leaves reserved for basements and servants' quarters. The other problem is the distortion and shrinkage when they are placed in a centrally heated environment.

6.5 Flush doors

The use of flush doors is found in all types of buildings.

The popularity is due to their proven reliability and to the fact that designs can accommodate a wide range of performance standards, for example, fire resistance, glazing and sound control. There is a wide choice of finish, from paint or spray coating on hardboard to clear lacquers on selected veneers. Laminate plastic and metal facings are available together with moulded facing 'skins' that simulate panelled designs. Suppliers offer film wrapping that can be retained to give site protection.

The basic forms for internal use are described in BS 4786 : Part 1, Fig. 6.7 illustrates the sizes as well as the patterns of glazed panels that can be provided. Imperial sizes may also be obtained from most suppliers and are indicated in brackets. With modern manufacturing techniques, alternative facing 'skins' can be matched to various core materials, so that, for example, fire and non-fire doors can look identical. The cores can be solid, part-solid or cellular, those without solid cores need blocking pieces to enable secure screw fixings to be made for hinges, latches/locks and closers (Fig. 6.8). Many suppliers have standard locations — the wise designer establishes the blocking to suit ironmongery and insurer's specification. There are the same requirements for door strengths as DD 171 (see Table 6.2). The governing factors in the construction are the type and weight of core material used and how framing methods are employed. See Fig. 6.7 for core and framing, noting that the weights given for comparison purposes relate to a door size of 826 × 2040 mm.

The specification for flush doors is also governed by the following recommendations:

- BS 1186 : Part 1 and BS 6566 : Parts 1−8 give directions on timber and plywood respectively. The Standards describe the quality of timber or ply, moisture content, amount of acceptable sapwood, freedom from decay and insect attack, limitation of checks and splits and treatment of resin staining, and the way plugging may be employed to mask defects in ply faces. Clearly higher standards will be needed where clear finishing is to be employed. British Standards permit pin-worm holes under some circumstances, guidance on this aspect being given in the *Forest Products Research Leaflet* No.17. In high-quality work it is worthwhile writing specifications that preclude any timber with pin-worm holes to save confusion. The direction of the ply grain should be vertical for paint or for clear finishes. Lippings if asked for, should be at least 7 mm in thickness, but 12 mm will give better service.
- BS 1141 describes hardboard facings while BS 5442 : Part 3 : 1979 and BS 1204 cover adhesives for use with wood with type MR or preferably type WPB for robust situations.

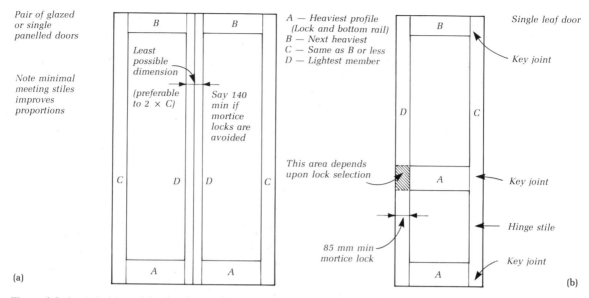

Figure 6.6 Logical sizing of framing for panelled doors (a) and (b).

- BS 1186 : Part 2 : 1971 governs workmanship.

Methods of manufacture and unit costs vary considerably; some makers serve the low-price market and can offer no variation from the standard products, while others offer a design and fabrication service to meet performance specifications. Door, jamb, head, fanlight and sidelights, glazing and ironmongery are provided as a package. A visit to both types of factory process is illuminating and will explain the price range from £15 to £150 for a single door leaf. The crucial aspects which affect the cost are core construction, type of framing, facings and lippings:

- *Cellular or lattice core* Used for the cheapest construction where a cellular core is made as a hardboard 'eggcrate' or lattice (as Figs 6.9 and 6.10). Other lightweight cores include spiral paper or wood coils and expanded cellular paper. The thicknesses are 40 and 44 mm with weight from 15 to 21 kg for light-duty application and designated grade III by the BRE for the PSA and DHSS.[2]
- *Skeleton frame* Traditional method of framing with battens and rails forming horizontal infill between periphery frame as Fig. 6.11(a). Weight and thickness governs grade: 40 mm thickness, with 30−40 percent core timber and 16 kg weight achieves light duty, and grade III (BRE) 44 mm thickness, 50−60 percent core timber and 20 kg weight achieves medium duty and grade II (BRE). This method is still used by small

joinery work in the making of non-standard doors, though seldom employed in large-scale manufacture.
- *Semi-solid construction* Timber framing increased to 50−60 percent of core (Fig. 6.11(b)) or with infilling high density flaxboard (weight 35 kg) or extruded cored chipboard (weight 32 kg). The construction can achieve medium duty and grade II (BRE) and is usually ply faced.
- *Solid core doors* The laminated core is formed with selected softwood battens glued together vertically to form a solid background to the ply (as Fig. 6.12). This gives the strongest form of construction and will withstand heavy use over a long period. The minimum thickness of 44 mm weighs 39 kg. Other forms of solid cores are made as follows: *wood chipboard*, with 44 mm thickness, weighing 45 kg; *plasterboard and cement fibre*, laid within skeleton framing to form solid core, 44−55 mm thickness, weighing 39−62 kg.

In the case of high-quality construction, the batten core is framed from western red cedar to provide a timber with minimal movement, the timber grain being laid in alternate and opposite directions to balance stress and thus reduce distortion. Ply facings are used with horizontal layers behind vertical face veneers that would be carefully matched for a suite of doors. Hardwood lippings for this class of work are run to four edges. Door blanks that incorporate laminated cores are made in large sizes and are useful in refurbishment work where the door is cut to the size of the opening, with lippings applied on site.

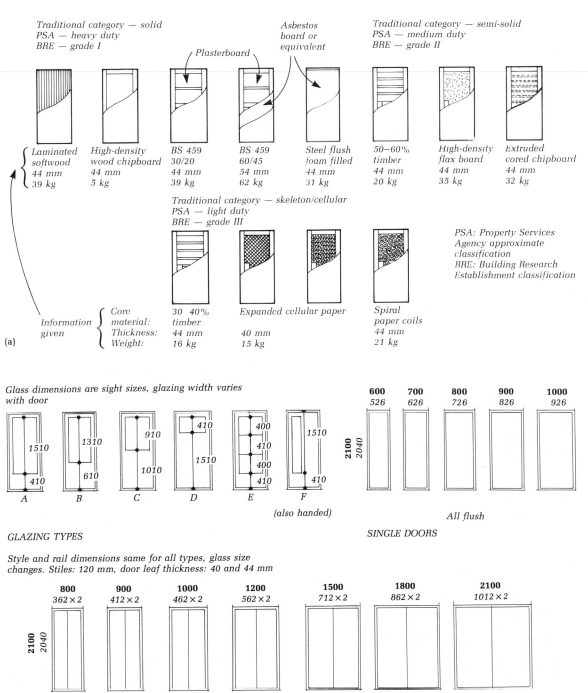

(a)

Traditional category — solid
PSA — heavy duty
BRE — grade I

Plasterboard

Asbestos board or equivalent

Traditional category — semi-solid
PSA — medium duty
BRE — grade II

Laminated softwood
44 mm
39 kg

High-density wood chipboard
44 mm
5 kg

BS 459 30/20
44 mm
39 kg

BS 459 60/45
54 mm
62 kg

Steel flush foam filled
44 mm
31 kg

50–60% timber
44 mm
20 kg

High-density flax board
44 mm
35 kg

Extruded cored chipboard
44 mm
32 kg

Traditional category — skeleton/cellular
PSA — light duty
BRE — grade III

PSA: Property Services Agency approximate classification
BRE: Building Research Establishment classification

Information given
Core material: timber   Thickness: Weight:

30–40% timber
44 mm
16 kg

Expanded cellular paper
40 mm
15 kg

Spiral paper coils
44 mm
21 kg

Glass dimensions are sight sizes, glazing width varies with door

1510   410
1310   610
910   1010
410   1510
400 410 400 410
1510 410

A   B   C   D   E   F
(also handed)

GLAZING TYPES

600 / 526   700 / 626   800 / 726   900 / 826   1000 / 926

2100 / 2040

All flush

SINGLE DOORS

Style and rail dimensions same for all types, glass size changes. Stiles: 120 mm, door leaf thickness: 40 and 44 mm

800 / 362×2   900 / 412×2   1000 / 462×2   1200 / 562×2   1500 / 712×2   1800 / 862×2   2100 / 1012×2

2100 / 2040

All flush

(b)   DOUBLE DOORS (add 6 mm each leaf for 12 mm rebate)

Figure 6.7 Internal flush doors to BS 4786 : Part 1: (a) core construction; (b) elevations — dimensions: **co-ord** & *door leaf*. (By kind permission of *The Architects' Journal*.)

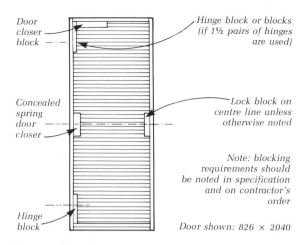

Door closer block

Hinge block or blocks (if 1½ pairs of hinges are used)

Concealed spring door closer

Lock block on centre line unless otherwise noted

Note: blocking requirements should be noted in specification and on contractor's order

Hinge block

Door shown: 826 × 2040

Figure 6.8 Possible blocking requirements for cellular core flush doors.

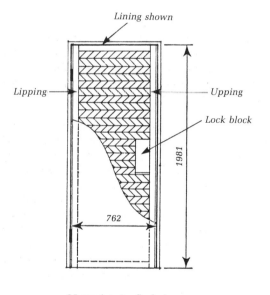

Lining shown

Lipping

Upping

Lock block

1981

762

35 mm interior flush door — expanded cellular board infill

Figure 6.9 Cellular core using expanded board infill.

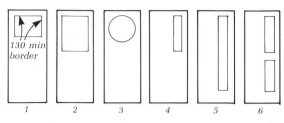

130 min border

Examples of glazed apertures in fire doors

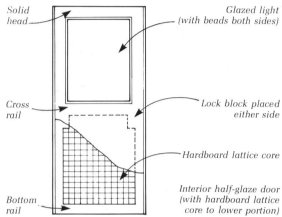

Solid head

Glazed light (with beads both sides)

Cross rail

Lock block placed either side

Hardboard lattice core

Interior half-glaze door (with hardboard lattice core to lower portion)

Bottom rail

Figure 6.10 Hardboard lattice core.

background for paint although the grain is usually revealed.

- *Selected ply facings*   These will suit factory or site application of clear finishes and are customary for grade I and II doors (BRE).
- *Grain printing on hardboard or ply*   This process utilizes furniture techniques whereby hardboard or ply acquires a simulated texture, either hardwood figuring or a panelled texture. They are available on all door grades.
- *Laminates, metal facing and hardwood veneer*   These expensive techniques are applied to grade I doors (BRE) and imply special productions. It is possible to apply laminates and veneers to all edges though damage will be difficult to rectify as compared to using face lippings in the conventional manner (see Fig. 6.4).

6.5.2 Protection of flush doors

The specifier needs to take regard of site protection; firstly, that dry storage is given within the building, with doors stacked horizontally on at least three cross-bearers. Stacking doors on their edge in damp or dry conditions will transform door leaves into banana shapes. Pre-hung door sets can save a lot of trouble and the provision of loose pin hinges means

6.5.1 Facings to doors

The cost for facing flush doors is given in ascending order:

- *Hardboard*   This is the cheapest but provides an excellent background for paint finishes. Protection of long edges with softwood lipping will improve performance of the paint at arrises and permit neater fixing of hinges and locks.
- *Cheap ply facings*   These will also provide a

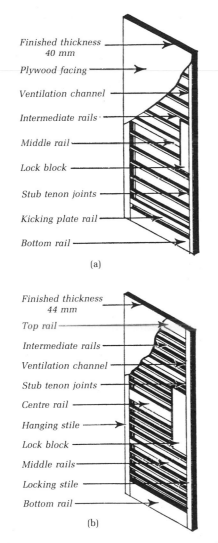

Finished thickness 40 mm
Plywood facing
Ventilation channel
Intermediate rails
Middle rail
Lock block
Stub tenon joints
Kicking plate rail
Bottom rail

(a)

Finished thickness 44 mm
Top rail
Intermediate rails
Ventilation channel
Stub tenon joints
Centre rail
Hanging stile
Lock block
Middle rails
Locking stile
Bottom rail

(b)

Figure 6.11 Skeleton frames for (a) light and (b) medium duty.

that installation can proceed with the opportunity to store doors safely away from the ravages of the site until handover.

6.6 Door frames and linings

Doors are hung on either frames or linings within an opening or screen. Linings are designed to face the reveals (sides) and soffit (head) of an opening and have to be adequately secured to support the door hinges and to bear the brunt of latches being slammed. Frames are of stouter construction, often rebated, say 95 × 45 mm minimum profile and employed where heavy doors are hung or to give greater security against break-in or fire. In either case

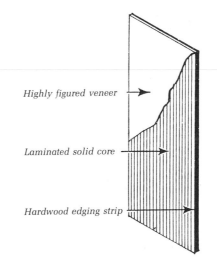

Highly figured veneer
Laminated solid core
Hardwood edging strip

Figure 6.12 Solid core door.

the linings and frames should be so installed to avoid loading from the masonry structure above the opening. A 6 mm gap is customary to allow for tolerances, the exception would be timber-framed construction where glazed screen framing is formed from wrot studwork and where doors could be hung directly. Wrot linings would of course be used for sawn studwork partitions, and in these circumstances it is a good precaution to install head trimmers with a separation of 6 mm so that the loadbearing wooden lintel can settle independent of the door head. Linings and frames should be fully primed before fixing to prevent absorption of moisture from the surroundings and from plaster. A common mistake on site is to build in lining or frame horns to obtain a better fixing in blockwork masonry. This inbuilt bearing will of course act partly as a lintel support which transfers hanging or slamming stresses to the lintel and leads to cracking of the masonry above (Fig. 6.13(a)).

The other critical points in lining and frame design are numerated as given under the subheadings below.

6.6.1 British Standards

BS 1567 : 1953 covers linings and frames for low-cost work. The 45 mm rebated frames are far superior since they enable the timber profile to clip blockwork masonry where this is built up with the joinery (Fig. 6.13(b)); the other advantage is sufficient purchase for hinge screws. The solidity of the 45 mm timbers gives a perfect rule for the plasterer's finishing work. Thin linings suit studwork, partitions and conversion work but the 22 mm thickness will need increasing to 32 mm for medium-duty doors (Fig.

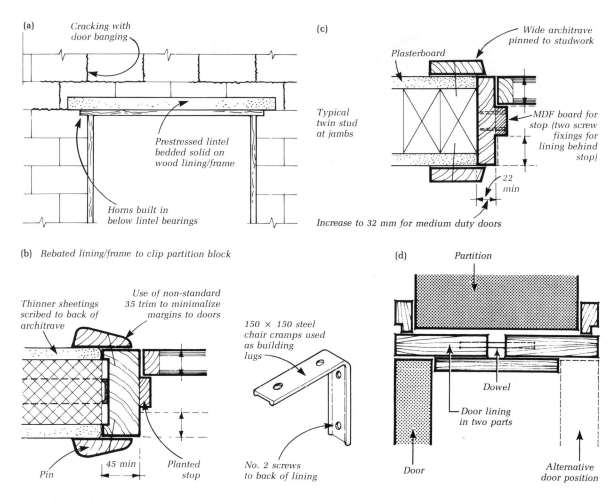

Figure 6.13 Linings and frames: (a) mistake with built-in frames and linings; (b) door frames to clip partition blocks; (c) thin linings for use with studwork; (d) two-piece lining.

6.13(c)). Planted stops of 12 mm thickness are supplied for pinning and gluing to linings to save the volume of timber utilized. Wide girth linings say 225 mm and over, can be made with 25 mm MDF board. The make-up size for linings should allow 2–3 mm clearance (in total) for hanging and adjusting the door. Stout doors of 60 mm thickness and above often have splayed edges to give a better fit.

6.6.2 Two-piece linings

High-quality joinery will comprise door sets where linings and doors are supplied complete with ironmongery and designed for universal application. The dowelled pair of linings can be adjusted to suit varying partition thicknesses and the planted stop machined to sizes to enable doors to

be hung to either face of the opening (Fig. 6.13(d)). A further refinement can be made by building in rough linings which in turn become fixing grounds when the pre-decorated linings and door sets are ready for installation (Fig. 6.14).

6.6.3 Built-in versus fixed-in joinery

The basic difference is a matter of labour cost. Linings and frames built into blockwork and masonry work act as a guide for the following block or bricklayers. It is essential that the linings and frames are stayed and have struts to keep the opening square. Back cramps are provided for building into the surrounding wall at 450 mm intervals starting 225 mm above floor level. Rough linings can be fixed on the same basis. Fixed-in joinery will need screwing

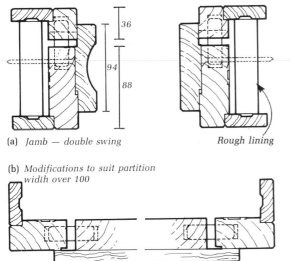

Figure 6.14   Rough linings and door sets. (Detail reprinted by kind permission of N. T. Shapland and Petter Ltd.)

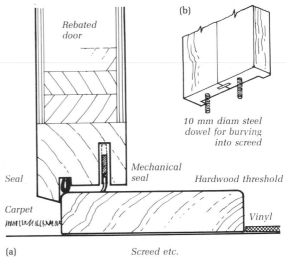

Figure 6.15   Threshold details: (a) raised hardwood threshold and rebated joinery; (b) dowel fixings for frames.

into prepared fixings (plugs or 10 mm pieces of ply built into masonry joints) and is more labour-intensive, both in terms of preparation and with facing up screw holes unless wide stops are employed to cover the fixings.

6.6.4 Thresholds and dowel joints

Hardwood threshold timbers are commonly used in high-quality work and have the advantage that differing floor finishes and thicknesses can be accommodated at door openings. Double rebated door edges can also be detailed at the face of the opening where greater standards of airtightness or sound deadening are needed (Fig. 6.15(a)). Raised thresholds have a considerable advantage in improvement work where raised floors may be installed above existing flooring for electronic services or else to improve sound insulation between flat conversions. Large profile linings or frames will need bottom fixings to reinforce the joinery against loosening, a simple device is to use a 10 mm steel dowel that is housed into the woodwork and cast into the floor screed or glued into blocking pieces (Fig. 6.15(b)). Raised thresholds, though used in domestic work in the UK, do not feature in commercial buildings due to the impediment caused.

6.7 Standard trim

BS 584 : 1980 describes standard profiles that are suitable for low-cost work. The architraves are shown in Fig. 6.16(a). Traditional patterns not covered by British

Standards are illustrated from part of Travers—Perkins catalogue. Superior designs are obtainable with door sets (see Fig. 6.14) where back-grooved trim permits a better fit to be made between wall finish and lining or frame. Purpose-made joinery trim can be afforded on large contracts but is a luxury item with small-scale contracts like the one-off house.

Standard profiles for architraves and skirtings allow for selection that gives a thicker section for the architrave, thus enabling this to master a thinner skirting profile. Expensive traditional detailing overcame these problems with architrave blocks glued and dowelled to skirting and architrave (compare Figs 6.16(c) and (d)). The use of standard door stop material is a cheap alternative; another device is to dispense with trim altogether and to infill with a gasket between lining/frame and the masonry surround (Fig. 6.16(e)).

Full-size drawings of the complete assembly — door, lining, architrave, plaster and masonry opening — are crucial in arriving at passage widths to accommodate standard doors. Figure 6.17 illustrates the way passages may have to be made 1 m wide between masonry faces to receive metric door sets and architraves; by comparison the imperial door size 762 mm (2ft 6in.) can fit within masonry spaced at 914 mm.

Joinery trim should have a moisture content of 14 percent and be primed before fixing. It is equally important to seal the back with primer or varnish if clear finishing is to be the final decoration. Penetration of moisture from drying masonry or plaster staining gives considerable problems.

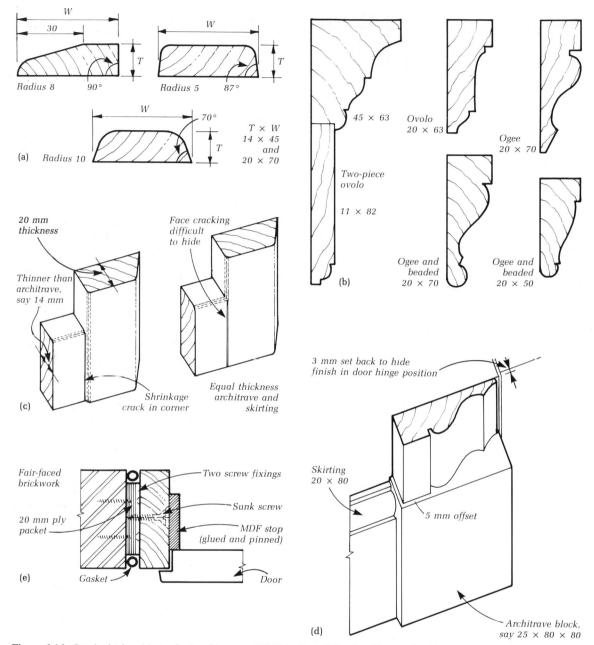

Figure 6.16 Standard trim: (a) standard architrave to BS 584; (b) traditional architraves (by kind permission of Travers Perkins plc); (c) architrave to skirting junction; (d) architrave block; (e) gasket used as a gap filler between linings and frames and masonry.

6.8 Metal linings, frames and trim

BS 1245 : 1986 describes standard pressed steel profiles for linings, frames and skirtings. The technology involves 18-gauge galvanized steel folded to shape so that rebates and architraves are in one piece (Fig. 6.18(a)), the

elevational arrangements allowing for single and double doors and fanlights. Temporary or permanent threshold pieces are provided. Improvements to the standard ranges include plastic clip-on architraves, draft stripping and pre-painted finishes. Door sets are also made with pressed steel or timber door leaves (Fig. 6.18(b)). Metal linings can be

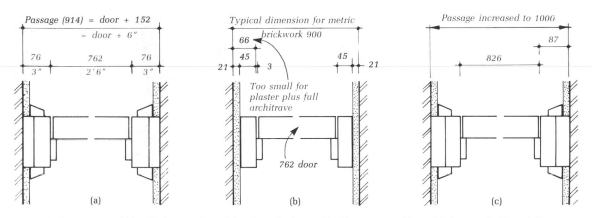

Figure 6.17 Passage widths. Setting out imperial and metric doors: (a) older pattern of imperial door set in 'imperial' corridor; (b) trying to fit an imperial door into a 'metric' corridor; (c) wider corridor resulting from all metric dimensions.

noisy in use but provide a stable, sturdy trim to doors in industrial and public buildings.

British Standard linings are designed for building in advance of partitions while the improved versions such as Fig. 6.18(b) are secured to rough linings at a later stage.

6.9 Fire-resisting doors

Fire-resisting doors need to be understood in terms of the whole framework of legislation that governs compartment zones of buildings and the provision of means of escape in case of fire. The standards of construction for fire doors (and their integrity) and for frames and screens have been refined in recent years and which are embodied in the new revised Building Regulations which came into force on 1 June 1992. A significant change in new technology permits doors to be assessed both from the point of smoke resistance and integrity under fire testing. The coding classification for fire doors is explained as follows. For example, a coding might be FD20 or FD20s. Here FD stands for fire door and 20 means 20 minutes where integrity is maintained under test applied separately from either side. The time length can be 30, 60, 120 and 240 minutes, identified by colour-coded plugs,[3] those at 30 rating and above being called *fire-resisting doors*. Finally, 's' is where smoke resistance is needed at ambient temperatures. The leakage rate should not exceed 3 m³/m per hour (head and jambs only) when tested to 25 Pa to BS 476 section 31.1 unless pressurization techniques complying with BS 5588 : Part 4 are used.

A review of the topical British Standards is given below.

6.9.1 British Standards

PD 6512 : Part 1 : 1985 *Guide to fire doors*. BS 588 *Fire precautions in the design and construction of buildings*.

This pair of documents need to be understood as they set out the reasons behind the provision and performance criteria.

PD 6512 : Part 3 : 1987 *Guide to fire performance of glass*. Look also at amendments, particularly revisions to BS 476 : Parts 20−22 that relate to glazed screens. See Fig. 6.20 for areas where requirement N1 of the Building Regulations applies.

BS 476 : Part 31 : 1983 *Methods of measuring smoke penetration through doorsets and shutter assemblies*. Section 31.1 deals with measurement under ambient temperatures, and sections 31.2 and 32.2 cover test for smoke at medium and high temperatures.

BS 5588 *Fire precautions in the design and construction of buildings*. One of the most important changes is replacing Code of Practice 3. The subdivisions are cross-referenced with the Building Regulations (being updated) and relate to varying building types:

Part 1 : section 1.1 1984 *Single family dwelling-houses*.
Part 2 : 1985 *Shops*.
Part 3 : 1983 *Office buildings*.
Part 4 : 1978 *Smoke control in protected escape routes using pressurization*.
Part 5 : 1986 *Fire fighting stairway and lifts (including fire doors)*.
Part 8 : 1988 *Escape for the disabled*.

BS 4787 : Part 1 : 1985 *Doorsets, door leaves and frames*. Defines sizes but does not include performance grades.

BS 8214 : 1990 *Fire door assemblies with non-metallic leaves*.

6.9.2 Functional requirements for fire doors

PD 6512, already referred to as the key document, makes new definitions for fire doors. Firstly, to protect escape

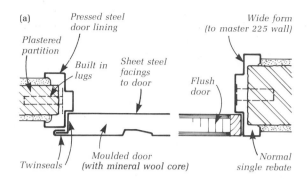

(a)

Plastered partition — Pressed steel door lining — Built in lugs — Sheet steel facings to door — Twinseals — Moulded door (with mineral wool core) — Wide form (to master 225 wall) — Flush door — Normal single rebate

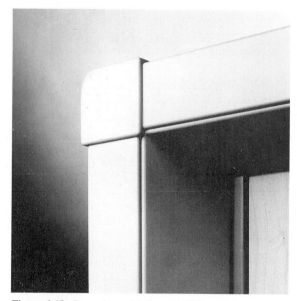

(b)

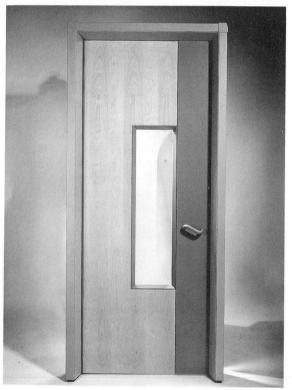

Figure 6.18 Pressed metal linings and frames: (a) typical lining; (b) metal door lining and door set. (By kind permission of G. & S. Allgood and the designer Alan Tye Design RDI Ltd.)

routes from the effect of fire so that the occupants can safely reach the final exit to the open air or ground floor. Secondly, to protect the contents and/or structure by limiting fire spread. The function of a fire door has to provide one or both functions in the following situations — smoke control, means of escape and compartmentation.

Smoke control This aspect receives greater attention than previously and point the way to using flexible seals plus intumescent seals instead of 25 mm rebates and intumescent seals. The new technology involves nylon 'brush' seals or inset synthetic rubber pads which give an effective seal against smoke in the early stages of fire. The intumescent seals foam as the temperature rises and close off the gap at door edges. Door makers and fire officers are not in agreement on these matters. The traditional 25 × 45 mm screwed and glued fire stop provides a stiffener to slender linings while the solid rebated hardwood frame (70 × 95 with 25 mm rebates) has a proven track record. These traditional means coupled with intumescent strips are still preferred by many specialists. The robustness of the door construction is also important to prevent warping and smoke leakage (see Fig. 6.29 for application of seal).

Escape doors   These have to hold back smoke but also be able to maintain integrity during a fire for sufficient time to permit occupants to escape.

Door categories   The new Parts 20/22 of BS 476 stipulate three categories as scheduled:

1. *Smoke control doors*   Graded according to ambient temperatures of smoke.
2. *Dual purpose smoke control and means of escape doors*   As item 1 but integrity of 20 minutes.
3. *Compartment and high-risk doors*   As item 2, but integrity of 30, 60, 120 or 240 minutes. High risk doors have the same fire resistance as the adjoining compartment wall.

Door sets   The new codes are specific about the fire resistance of the door, the frame and seals, glazed elements and the connection to the enclosing wall. One of the advantages of installing door sets is the guarantee that is available. Makers experienced in putting together door sets are used to performance specification and have testing facilities which ensure compliance with regulations.[4] Wood or metal frames which are to be fixed by the builder will require careful specification, namely full and proper fixings to prevent distortion under heat stress and packing with mineral fibre, mortar or concrete to absorb heat flow at the gap between the framing and the surrounding building structure.

The detailed construction of fire-resisting doors varies according to the supplier, for example timber flush doors with plasterboard, cement fibre or mineral wool cores, the faces being ply, or moulded board. Steel-clad and framed doors infilled with fibre-reinforced gypsum or doors infilled with mineral fibre can provide unit ratings to meet FD60, 120 and 240. Heavy doors are often made as sliders (see section 6.10).

The 'burn rate' for wood flush doors can be reduced by constructing in hardwood with a density in excess of 650 kg/m^3 and is utilized in solid core doors of 44 mm thickness made to FD60 rating.

6.9.3 Ironmongery

The selection and installation have to be agreed with the fire officers and insurers; the significant details are illustrated in Fig. 6.19. The requirements for fire testing are laid down in BS 476 : 1972 and where the door specimen and frame have to be tested with essential ironmongery in place, care is needed to ensure that morticing of hinges and locks does not reduce the fire stopping of the doors. Today intumescent material in the form of plugs are packed behind hinge, latch and striking plate positions with a paste sealant applied round latches and locks, etc. Self-closing devices are an essential feature for all fire doors, including smoke control. They should be able to close a door fully and to engage the latch, the mounting position always being on the safe side of a fire door to limit heat damage. In less hazardous conditions, such as housing, rising butts are permitted (steel with brass bushes). Double doors may require rebated meeting stiles and have to be fitted with a door selector to ensure proper closing, the selector having sufficient strength to resist the pressure exerted by fire stresses upon door enclosures (see Fig. 7.19). Further advice is available from the Association of Builders Hardware Manufacturers who publish a Code of Practice for hardware essential to the optimum performance of emergency exits and fire-resisting doors; an allied document can be obtained from the Guild of Architectural Ironmongers. A potential conflict exists in selecting escape door furniture, namely ease of escape in case of fire and the need to provide security against unlawful entry. Further information is given in Chapter 7 (ironmongery).

6.9.4. Glazing of fire doors and screens

Glazing has a number of advantages in fire doors. Firstly the ability to see if an escape route is clear. It also assists staff in clearing a building in case of emergency, since vision panels enable rooms to be checked without opening a fire door.

Secondly, the advantage in fire-fighting when a hose can be pushed through a hole cut in the glazing. This enables the fireman to quench the fire in relative safety.

PD 6512 : Part 3 gives guidance on fire performance of glazing but lacks specific advice, the details in Fig. 6.20 being taken from makers' recommendations for glazed doors and screens.

6.10 Sliding fire doors and vertical shutters

Where fire doors are required to be rated FD60, FD120 and FD240 they are often metal framed and protected by glass-reinforced gypsum cores. The FD120 and 240 rated doors are heavy to close and are therefore hung on an inclined overhead track. The door will normally be held open by a counter-weight connected to a fusible link. The

Top edge of door eased
if rising butts used
and deeper stop required

*Intumescent
strip*

*Hinge fixings must ensure hinge
holds seven when severe charring
occurs. Hinge may interrupt
intumescent strip — check that
this does not negate the test
certificate*

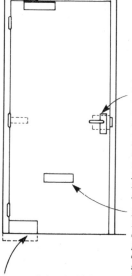

*Proper maintenance of door
closers is essential on fire doors
(inform clients)*

*Recessed door closer can
seriously lessen fire resistance of
door — discuss with manufacturer*

*Door latch/lock — fill voids in
cut-out for mortice lock with
intumescent paste, nib of
latch/lock should engage latch
plate (strike plate) by min
10 mm. ER 60/60 doors. Rim
locks should be mounted on non-
risk side of door or be bolted
through thickness of door. Door
furniture — material used is not
normally critical but spindles
should be steel*

*Letter plate — should be avoided
if possible, if essential, e.g. flat
entrance door, locate in lower
part of door. Flaps are required
on both faces, ideally both in
steel but at least inner flap must
be steel*

Recessed threshold door
closer is best from fire
resistance point of view

Figure 6.19 Installation of ironmongery on fire doors.

link breaks when a critical temperature is reached, allowing the door to slide shut. Electromagnetic devices can also be used that are wired to heat sensors within the fire compartment. Figure 6.21 illustrates the conventional installation with fusible link operation. Other factors are given under the subheadings below.

6.10.1 Air permeability

Sliding doors are available with plastic trims to give restrictions on air movement; these could be useful with industrial processes where airborne dust is a nuisance. The trims do not affect fire resistance because the edge rebates into which the door runs are 150 mm or more and of steel construction.

6.10.2 Cleanliness

With horizontal sliding doors, the major difficulty is the bottom guiding track which is very difficult to clean, if in channel form. The vertical sliding pattern is therefore preferred in food-preparation areas (either by a one-piece or folding operation). Doors for areas where hygiene is important can be enamelled or faced with laminate or vinyl by most manufacturers.

6.10.3 Operational factors

Size (horizontal operation). There are no real limitations (except self-weight) and horizontal sliding doors can be made up 10–12 m wide. The time taken to operate is a factor, and the designer may then prefer pairs of doors or leafing on different tracks so that the open doors are stacked within a short 'parking bay'. The height limit without a bottom track is 6 m and the elevational proportion should be at least 1:1.3 (height to width) to prevent jamming on top guides.

Bottom guides will be required for large sliders (height up to 10 m) and these should be recessed channels, especially where fork-lift trucks are running, to avoid tripping and damage.

6.10.4 Vertical sliders

Once again there is no limitation other than space over the opening and the deadweight of the single leaf. However, sizes over 4.5 m high are unusual. See Fig. 6.21(b) for arrangement.

6.10.5 Selection of door pattern

The specifier should look at the choice offered by industrial door manufacturers before deciding upon simple sliding operation, because recent developments with horizontal or vertical folding doors may provide solutions that are more economic in space requirements.

6.11 Roller shutter doors

6.11.1 Fire and smoke

Steel roller shutters are used as fire checks in many buildings, for example as compartment walls in department stores or factories. The closing action is triggered by an electrical alarm or fusible link, as for sliding doors. Sometimes a klaxon is used to give a warning when the shutter descends.

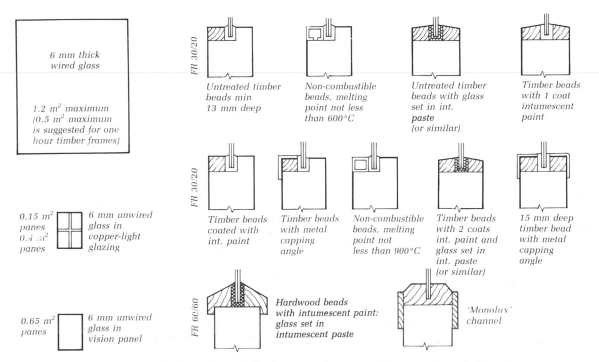

Figure 6.20 Recommendations for fire-resistant glazing in doors and screens, and impact-resistant glazing.

Steel construction for fire resistance The construction is usually galvanized steel with enamel finishes used where better appearance is required. Aluminium, plastic and timber look better but do not meet fire standards. It is not possible to insert glazed elements.

Insurance and Fire Offices Committee requirements The Fire Offices Committee (FOC) lay down that the structural opening shall not be more than 2.4 m wide and 2.1 m high for self-coiling roller shutters, with a maximum area of 5.04 m². Larger sizes are feasible outside the rules of the FOC and where mechanical operation is installed. The standards for fire shutters are 2- and 4-hour fire resistance.

6.11.2 Poor noise reduction

Roller shutters are not normally regarded as providing any significant noise reduction, and are quite noisy themselves in operation.

6.11.3 Air permeability

The general arrangement with interleaving blades running in U-shaped guides gives excellent standards for air

permeability, but no figures are available from manufacturers to make comparisons with sliding doors.

6.11.4 Cleanliness

Enamelled finishes applied over galvanizing will give a 'wipe clean' standard that compares well with other industrial doors. If fire resistance is not important then there is a choice of anodized aluminium, plastic or stainless steel. These are of lighter construction and far easier to operate.

6.11.5 Operational factors

Disadvantages:

- Frequent maintenance (three-monthly oiling).
- Noisy in operation.
- Slow manual operation.
- Limited sizes (if working to FOC requirements) and manufacturer's limits.
- Difficult to adapt to automatic opening and closing for production processes.

Advantages:

- The closing operation under fire test conditions is fairly

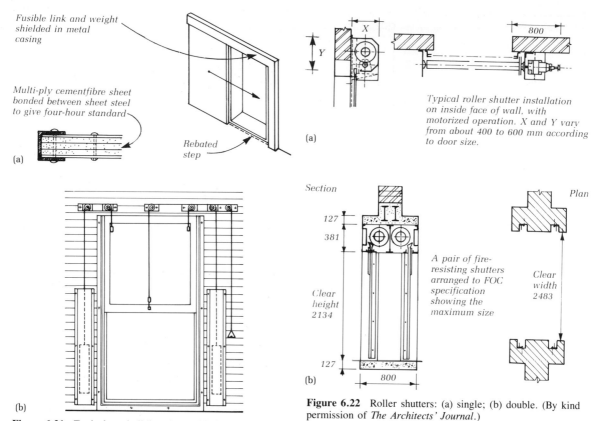

Fusible link and weight shielded in metal casing

Multi-ply cementfibre sheet bonded between sheet steel to give four-hour standard

Rebated step

(a)

(b)

Figure 6.21 Typical steel sliding door with fusible link operation: (a) horizontal pattern; (b) vertical pattern. (By kind permission of Cape Duraseel.)

Typical roller shutter installation on inside face of wall, with motorized operation. X and Y vary from about 400 to 600 mm according to door size.

(a)

Section   Plan

127

381

Clear height 2134

A pair of fire-resisting shutters arranged to FOC specification showing the maximum size

Clear width 2483

127

(b)   800

Figure 6.22 Roller shutters: (a) single; (b) double. (By kind permission of *The Architects' Journal*.)

slow and enables the occupants to step clear.

- The general robustness of the steel construction gives good security.
- The arrangement of the canopy, gearing and guides on one face of the component wall gives more uncluttered space than the sliding or sliding—folding format.

6.11.6 Points on installation

See Fig. 6.22(a) for typical 'face' installation. In shops it is customary to site the shutter within the wall thickness to save sales floor space. Where structures are not strong enough to support the rollers, portal-type frames can be used with columns taken from the base slab.

The lower edges to roller shutters can have special plates to follow the ground profile (e.g. where roads and kerbs pass through a fire door) and can also be arranged to match falls in floors or complex steps.

Bar counters can have detachable mullions that are locked into place so that a continuous face is given to the shutter

front, with timber slats (up to 2.4 m long) or timber-simulated steel to match the counter joinery. A wicket gate can be fitted which may have advantages in shop-front work.

6.12 Concertina and folding doors

These industrial-type doors are commonly used for lift gates and where automatic operation is essential.

6.12.1 Fire and smoke

Folding or sectional doors comprise narrow vertical panels (or pickets) connected by continuous vertical hinges and generally backed by a diagonal folding lattice framework, connected to verticals to prevent racking during operation. Electrical or manual control is possible: lift landing gates in industrial premises are a typical application. The pickets slide with a 'concertina' action to stack flat against one another when the door is open.

This format is probably the commonest type of industrial door. It is equally suitable for small or large openings and can provide a high degree of security and fire resistance (between 1 and 2 hours) The small size of picket, however, only allows small areas of glazing, and the panels cannot be thickened to provide high levels of thermal insulation.

6.12.2 Security

The same provisos exist as for roller shutters, with internal fixing behind the structure rather than face mounting to provide better protection against intruders.

6.12.3 Durability

The moving parts are larger than roller-blind slats and builders' paint finishes can be used. However, stove-enamelled coatings will give long life as will a bright metal finish for the lattice elements on smaller doors such as lift entries.

6.12.4 Noise reduction

Noise reduction is not really feasible owing to the thin plate construction of the pickets.

6.12.5 Air permeability

The general construction has similar qualities to roller shutters.

6.12.6 Cleanliness

The external shutter face can be well finished using stove-enamelled or vinyl-coated plate, the former being suitable for respraying to restore the original condition. The cheaper versions are galvanized sheet and must be finished with builders' paint, which is adequate for many industrial installations.

6.12.7 Operational factors

Disadvantages:

- Not possible to fit wicket gate.
- Fairly noisy in operation and not possible to insulate door panels.
- Needs 3-monthly greasing and maintenance.
- A bottom track 'U' or 'T' has to be fitted and both have serious disadvantages in terms of cleaning and damage under traffic conditions.

Advantages:

- Cheap installation.
- Takes up minimal space.

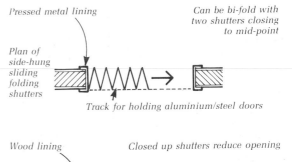

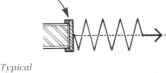

Figure 6.23 Typical plan layouts: (a) accordion or folding shutters plus photograph of installation; (b) 'Pella' accordion doors.

6.12.8 Points on installation

See Fig. 6.23 for typical plans. The limitation of size is 18 m high with no limit to widths, although very wide openings will need to be divided into 'floating' sections because friction in the hinged units makes long runs too heavy to move.

6.13 Collapsing gates or grilles

These open forms of grilles are designed for security against break-in and are employed for protecting bars or kiosks or for shop fronts in covered shopping malls (Fig. 6.24). Fire protection if needed will have to be provided with a back-up of fire shutters. The other points to consider are given under the subheadings below.

6.13.1 Durability

Site painting will seriously diminish the ease of operation, and designers should consider self-finished materials such as anodized aluminium, plastic (antistatic), stainless steel or coated steel.

Cement-based paving materials are corrosive to aluminium and it would be wise to revert to stainless steel for the bottom section in externally sited grilles.

Vandalism is a more serious problem in the open street than in a covered mall, and metal grilles are the best deterrent in a hostile environment.

Figure 6.24 Typical collapsing grilles.

Attractive patterns are made from clear materials such as polycarbonate which provides limited resistance to fire attack.

6.13.2 Air permeability

Open grille treatment in covered shopping centres means that the shop interior and mall environment are connected and that air curtains, which are expensive to run, will have to be used to separate heat zones or food smells.

Everything is possible if running costs are not the prime consideration, one of the olfactory miracles being at Valizy Deux in Paris, where the cheese emporium and fresh-fish trade were next door to one another without a hint of either at each other's counter.

6.13.3 Cleanliness

This aspect is linked to the choice of self-finished materials (see section 6.13.1). In food shops the working mechanism needs to be encased and to be totally removable for thorough cleaning and maintenance.

6.13.4 Operational factors

Operational factors are similar to those described for roller shutters. Traditionally, collapsible gates (like open-lift cage doors) have been used for security, this being a long-established method of protecting openings based on the 'lazy tongs' principle. The gates comprise steel uprights connected by a lattice of diagonal bracing which slide and fold horizontally on top and bottom runners. Smaller models

for protecting entrance doors are available as lift-away units. This type of gate can provide a high degree of security at reasonable cost and can be used in conjunction with other types of door, for example flexible doors which on their own cannot provide adequate security in external openings.

6.14 Sound-resistant doors

Sound waves can travel from one room to another by *flanking transmission* along and through the fabric of the building, *impact transmission* by vibrating directly through the thickness of the separating construction, or by *airborne transmission* through gaps in any part of the construction (see *MBS: Introduction to Building* Chapter 7). Doors within frames often provide the easier path for sound transmission when compared to an imperforate wall of most forms. The reason rests with the lighter unit weight of a door compared with the surrounding wall plus the frequency of air gaps between door and door frame and threshold. These shortcomings explain the fact that the net sound resistance through a wall perforated by a door will be limited to about 7 dB rating above that of the door and frame construction as demonstrated in Fig. 6.25. Simple improvements can be effected by detailing with edge seals and by constructing door leaves with cellular cores as Fig. 6.26; this provides a sound reduction of 30 dB. To achieve this reduction the door must be hung as shown with sealing strips at the rebate of the frame, and the frame must be adequately secured to the walling, which should be at least 190 mm thick solid masonry to achieve the reduction indicated. The decibel reduction required to eliminate a normal conversation from one side of a construction to another would be 27 dB. This is an adjusted level difference represented by the door in its frame as shown, well fitted in the wall. The reduction figure of 27 dB takes into account the combination of, and the different decibel ratings for, door and wall.

One effective way of increasing the sound resistance of a wall requiring a door opening is to provide a pair of doors opposite to one another within the thickness of the wall construction. The jambs between the two doors and the inside face of the doors themselves should be lined with materials providing as much sound absorption as possible. Alternatively, as in sound recording studios, *sound-resisting lobbies* filled with absorption materials should be used to provide isolation.

6.15 Darkroom and X-ray room doors

Figure 6.27 indicates doors suitable for darkrooms, where photographic processes take place, and for resisting X-rays, as would be used for medical buildings.

6.16 Flexible doors

This type of door is used where it is not possible to open the door in the normal way. The usual applications are in industrial buildings, warehousing or hospitals where the user may be pushing or driving a trolley or carrying bulky packages. It is a comparatively inexpensive alternative to installing automatic opening and closing devices. The door is composed of a flexible membrane of either reinforced rubber or neoprene, or where complete vision is desired for safety in operation, transparent or translucent plastic sheet or strips. In any case, because of the risk of collision, this type of door would be fitted with a clear plastic vision panel.

The door is designed to open on impact, the flexibility of the sheeting taking the force out of the 'collision', and allowing the user to pass through. The doors then close automatically, being controlled by jamb spring hinges. The door illustrated in Fig. 6.28 is a lightweight door formed by a steel angle supporting frame at the head and hinged side, from which is hung a flexible sheet of 8 mm reinforced rubber clamped into the heel of the angle by means of a steel flat. The hinges are double action vertical spring type. For larger doors up to say 4.000 m high by 4.000 m wide

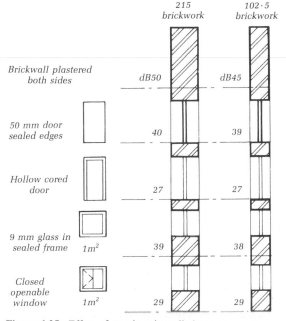

Figure 6.25 Effect of openings in walls have on sound attenuation performance. (By kind permission of the BDA.)

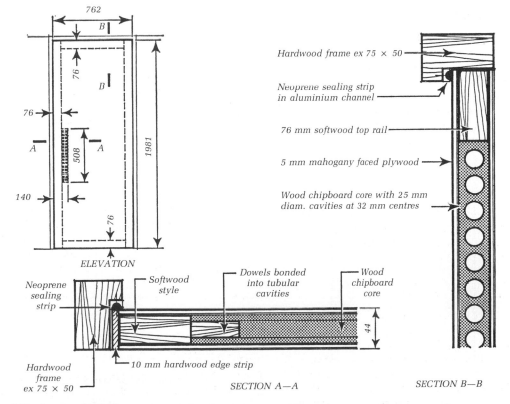

Figure 6.26 Sound-resistant door.

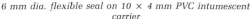

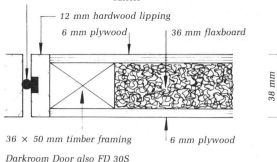

6 mm dia. flexible seal on 10 × 4 mm PVC intumescent carrier

12 mm hardwood lipping

6 mm plywood

36 mm flaxboard

38 mm

36 × 50 mm timber framing

6 mm plywood

Darkroom Door also FD 30S

2 mm lead sheet between two layers 6 mm plywood

12 Mm hardwood lipping

66 mm

36 × 50 mm timber framing

36 mm flaxboard

X-ray room Door also FD 30S

Intumescent seal in PVC carrier

Figure 6.27 Darkroom and X-ray room doors. (By kind permission of Castlecomer Woodwork Co. Ltd.)

the framing would be of steel tubing with the spring mechanism made as a separate detachable unit, and arranged to slide inside the vertical tube framing from the top. The heaviest doors are suitable for openings which are used by large vehicles and other heavy industrial transport.

6.17 Automatic control of doors

External and internal doors can be controlled in terms of opening and closing by use of various types of automatic equipment. A fully automated system will incorporate a device which acts as the initial sensing control such as a push-button, a sensitized mat or a photoelectric beam. This initial sensor will be followed by a timing device connected to the motor which causes the doors to move. Alternatively, the motor can be operated by remote control from a central point such as a security cabin. The timing apparatus can vary from a simple cut-out to complex electronic control with programmed instructions to incorporate variable time delays for closing and opening doors in series as circumstances or security require. It is essential that all automatic control devices allow the doors to be moved by hand in the event of a power failure. The motor gear usually takes its initial power from electricity which is then used to generate hydraulic or pneumatic pressure to cause the doors to move as signalled. The equipment should be capable of incorporating a checking action which slows the doors down towards closing in order to avoid clashing and rebound. A further refinement should be a repeat cycle of opening and closing should the door meet any obstruction.

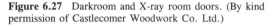

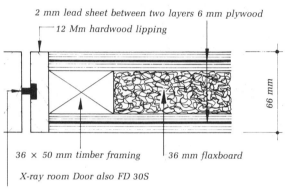

1·500

2·000

375 × 300 triangular viewing ports

Double-action spring butt hinge

ELEVATION

3 mm nom. m.s. flat panel retaining frame

Timber fillets

Timber door surround

8 mm nom. flexible fubber door panel

Double-action spring butt hinge

25 × 25 × 3 mm nom. m.s. equal angle door frame

HORIZONTAL SECTION AT JAMB

Figure 6.28 Lightweight flexible rubber doors.

ELEVATION

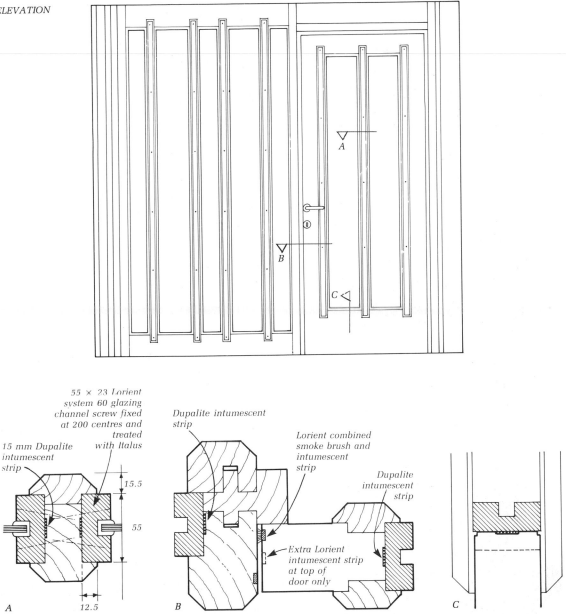

Figure 6.29 Fire screens and doors, Hall of Justice, Trinidad. Joinery and glazing door and screen details: section *A* — glazing stop, cover batten and muntin; section *B* — door edge, frame, stop and glazing plan; section *C* — section through base of glazing (door or screen). See section 6.9.2 for references to seals. (By kind permission of the architects, Anthony C Lewis partnership in association with Howell, Killick, Partridge and Amis.)

Specialist advice should be sought on this type of gear since automation techniques are rapidly developing, and more sophisticated control is possible.

Automatic control may be required in hospitals, hotels, shops, offices or security buildings, and both swing doors and sliding doors can be controlled. Briefly, for the swing door the control will be as follows, a master control with time delay for setting the time that the door is held open, and a regulating resistance for setting the speed of movement for the sliding doors. The leaves may be driven electromechanically by a driving wheel attached to a moving rail at the top of each leaf. The rail regulates the width of opening, brakes the door before the end of its movement, and makes electrical disconnection at the end of the movement. The maximum speed of opening for a sliding door will be in the region of 1 m per second.

6.18 Case Study: Fire screens and doors, Hall of Justice, Trinidad

Architects: Anthony C Lewis Partnership in association with Howell, Killick, Partridge and Amis.

The fire screens and doors occur between public areas and interview rooms in the courthouse and comprise various joinery assemblies (Fig. 6.29) that were tested in accordance with BS 476 Part 8 1972 to the following standards: stability — 62 minutes, integrity — 42 minutes; the glazed elements were classified as 30 or 60 minutes according to the size of glazing channel employed.

The door and screen details employ glazed strips set in grooved glazing channels placed over intumescent strips. These ensure that gaps between components are smoke-sealed as heat increases in the fire. The glazing is a multiple form with interleaves which in turn give higher insulation against the passage of heat during a fire. The cover battens are hardwood and rebated over the glazing channels to give extra security, this also provides a repetitive vertical feature to unite the design of doors and screens.

Notes

1   There are many detailing books emanating from Germany that are comprehensively illustrated. There are classic volumes titled *Detail* that are published details extracted from *Architektur + Baudetail*, either published by that magazine, with the earlier work printed by Iliffe Books Ltd in the 1960s. Full-size details for both doors and windows are available from textbooks by Adolf Schneck. Another source are published details in the *Architects' Journal* and the bound volumes called *Architects' Details Sheets* that are printed from time to time. For those with traditional tastes there are always the illustrated books by Nathaniel Lloyd which cover smaller Georgian houses. Finally, prewar copies of Mitchell's or Mackay's construction books are useful for those inspired by 'Bankers' or 'Brewers' Georgian'.

2   Refer to manufacturers' catalogues for precise weights and thicknesses. The guidance given in the text is based upon the technical notes prepared by John Duell in *Specification* (printed by MBC) and relates to a standard door size of 826 × 2040 mm. Table 1 for disc strengths is published by British Standards in their Draft for Development 171 : 1987.

3   Colour-coded plugs on the door edge indicate potential fire resistance rating when evaluated in accordance with BS 476 : Part B : 1972 or Parts 20/22 : 1987. The coding defines FD20 as well as FD30 and FD60, and describes whether intumescent material may be fitted at works or needs to be installed on site.

4   N.T. Shapland and Petter Ltd are typical of leading door manufacturers who belong to the TRADA QA scheme under BS 5750. They design, fabricate and install door sets complete with ironmongery to performance requirements.

7 Ironmongery

7.1. Introduction

Working components in buildings, such as windows and doors are required to hinge or pivot or else to fold and slide open. The mechanisms used for these moving parts and for the locking and opening devices are collectively called ironmongery, the older term being hardware. Window fittings have been included with internal ironmongery since hardware is normally chosen in a comprehensive manner, often as part of a total ironmongery schedule where timber windows and doors are involved. The fixings employed are an essential part of ironmongery and usually means bolts or screws supplied in the same finish as the ironmongery. The categories of items are summarized as follows: hinges and pivots; closing devices (hydraulic or spring); sliding track and their accessories; latches and locks; bolts and catches; knobs and handles; special fixings for cabinet construction and for coupling joinery.

It is essential that designers familiarize themselves with manufacturers' catalogues and realize the importance of the different qualities available, not forgetting that the best material will be the most expensive but costs less in the long term. The tactile quality is important where knobs and handles are involved, and it is worth while handling and looking at actual samples to study their application to doors and windows.

Insurance companies have required standards for locks and security fastenings and it is important to establish that the choice fulfils their requirements.

The choice of the correct type and quality of ironmongery is very important since, no matter how good the windows, doors and latches are, they will fail to function and cause annoyance if the working mechanisms controlling their movements are not durable and efficient.

7.2 Performance requirements

The performance requirements which ironmongery has to fulfil are formidable. For instance, the self-closing device of a door into a shop may operate hundreds of times every day and, in a department store, perhaps a million times a year. The door must be easy for a child or disabled person to push open, but must be capable of holding the door closed and not allowing it to be opened by the wind. During its working life there should be no breakage, it should be maintenance free and not susceptible to corrosion, and should be easily replaced when necessary. Working against these criteria lies the fact that ironmongery is one of the easiest and, therefore, the most likely item of expenditure to be cut back if building construction costs rise on a project. Nevertheless, cost-in-use is important, and specifiers need to satisfy their clients' increased awareness of financial outlay for the continual maintenance of buildings, particularly in the commercial sector.

Further comments on performance requirements will be included under each item of ironmongery considered.

7.3 Materials and finishes

The traditional materials for ironmongery have been cast iron, wrought iron, steel, brass and bronze. These are joined today by nickel-bronze alloys, stainless steel, aluminium alloys and plastic. The appropriate qualities are listed under the headings below.

7.3.1 Cast iron

This is used for garden gate and shed furniture and for decorative items like coffin nails and for fittings in 'Gothik' taste. Many British designs date back to the 1850s. It is available in black enamel finishes, resistant to corrosion but should be sherardized in exposed situations.

7.3.2 Wrought iron

A traditional material for all fittings up to the twentieth

century. Wrought iron needs protection by sherardizing externally where conservation of existing work is considered. It can be repaired by welding to mild steel in case of difficulties in finding supplies of wrought iron.

7.3.3 Steel

Most commonly employed for hinges and for lock cases and end plates to latches and locks, it is also used for all forms of bolt. It needs sherardizing or other zinc coating to overcome rusting problems. Cheap fittings are often steel-based with finishes such as enamel or coatings of chromium or bronze spray.

7.3.4 Stainless steel

Various grades are available for exterior and interior use. It has higher cost than coated steel but is far superior in terms of durability. The finishes available range from bright polished to matt surface. It is highly durable and therefore excellent for external applications (hinges, handles, door plates and bolts) and for damp/dirty situations within buildings.

7.3.5 Brass and bronze

Bronze has been used since Roman times and is an alloy of copper and tin. Brass casting dates from the sixteenth and seventeenth centuries and is made by alloying copper and zinc in 70:30 to 60:40 proportions depending upon the ductility needed. Both can be cast and forged and finished to a high standard, and often used as a base for plated finishes such as chromium or nickel plate. Natural coloured finishes include polishing and lacquer and which need careful maintenance. Lack of polish will mean that the materials oxidize and develop a green patina. Traditional ironmongery is available in brass and bronze; however, the majority of designs do convey the impression that the Bauhaus never existed. There are notable exceptions when designers like Alan Tye or Robert Welch are involved. Alloys such as nickel bronze are often retained where a silver/yellow colour is required, which benefits from polish or lacquer treatment.

7.3.6 Aluminium

The process for making aluminium was developed in the late nineteenth century. It has to be employed in alloy form (such as Duralumin) to give the requisite strength and surface toughness. The manufacture depends upon extrusion or die-casting and can provide a high quality of detail.

Finishes are applied by anodizing or coating. The anodic film may be coloured or polished or lacquered, and have a life externally of around 20–30 years. Aluminium alloys are reserved for knobs, handles, door plates, trim (lock plates) and minor items such as light hinges, door tracks and catches.

7.3.7 Plastic

Nylon is the commonest plastic used in the making of lightweight fittings, or simple latches, knobs and handles. The other application is coating to steel or aluminium cores. Moving or sliding components are often nylon coated or washered to provide a smooth, tough, wearing surface. The other application are washers to avoid bimetallic corrosion (e.g. steel fastenings to aluminium windows). The softening of plastics under fire exposure prevents their general use as hinge washers in fire doors and fire screens.

7.3.8 Metals and finishes on metals

For general points on metals and finishes see *MBS: Finishes* by Yvonne Dean (1989). Another useful guide is *Metals in the Service of Man*, by William Alexander and Arthur Street (Penguin Books 1989).

The popularity of aluminium alloys for ironmongery production stems from the low cost of the metal coupled with the low corrosion risk of the base material whether cast or extruded. The anodizing process will improve the external finish, producing an anti-corrosive film. The process involves electrodeposition with pure aluminium in a natural silver finish or else coloured with dyes (including bronze or gold). The life can be extended by lacquering. The surfaces, where regularly cleaned, will retain their brightness for a considerable period; from observation of pre-war buildings in Switzerland, certainly up to 50 years.

Comparative finishes such as chromium or nickel plating are applied over bronze or brass and will also provide long service if cleaned and polished. It is easier to strip and to replate after 40 or so years than to reanodize. Natural bronze finishes involve heating to give an iridescent quality that is termed 'bronze metal antique' (BMA). There are spray coatings that imitate BMA used in bronze, brass and steel but they are visually inferior. Polishing and waxing or lacquering are the other traditional finishes applied to bronze and brass ironmongery. Polishing will be needed frequently while waxing means a yearly sequence. Lacquering will need replenishing on a 5-year cycle. Modern finishes include powder coating that is fired on to the metal and can be applied generally to Duralumin and steel products and will last up to 20 years before recoating. Cheap iron and steel fittings are treated with enamel, termed 'Berlin black' or 'japanned laquer', which has poor durability, say, 2–3 years externally. Further consideration is given to the selection of knob and lever furniture in section 7.7.

7.4. Patterns of hinge

7.4.1 Differing types and their use

Butt The *butt hinge* is the most common hinge. It is screwed to the edge of a door and recessed into the frame as well as the door. Normally, one pair of 75 or 100 mm butts suits a standard internal door; external and other heavier doors (fire doors) might require one and a half pairs, i.e. three 100 mm butts. Butt hinges are made in steel (including stainless), brass with steel pins, brass with brass pins, aluminium or nylon, or ball-raced (Fig. 7.1).

Extended butts Extended butts are made for fold-upon-fold doors and for shutters. Washered butts will extend the life of external hinges and where heavy use occurs. Stainless washers or ball-bearing sleeves are the best patterns.

Piano hinge An extended version of butt hinge up to 2 m in length is termed a 'piano hinge' for use on keyboard covers, also used for wardrobe doors, usually made in brass.

Rising butts Rising butts lift the door as it opens so as to clear a carpeted floor, and this type of hinge is in some degree self-closing. Rising butts must be 'handed'. A falling butt hinge is also available and this will keep a door in the open position. Available in steel or brass with steel bearing plates.

Tee Tee hinges or cross-garnets are used for heavy doors of the ledged type (Fig. 7.2) available in steel. The construction can include ball sockets for strong service with heavy plates of lugs for securing at jambs (known as *Collinge* hinges.

Pin With pin hinges or lift-off butts, the door can be taken down without unscrewing the hinge and this type is, therefore, used for doors that are pre-hung and assembled in the factory (Fig. 7.3).

Back-flap Back-flap hinges are for screwing on the face of the work where the timber is too thin to screw into the edge, or where appearance is not important. They make a strong job when used on internal joinery fittings (Fig. 7.3). Available in steel or brass.

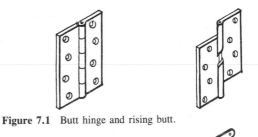

Figure 7.1 Butt hinge and rising butt.

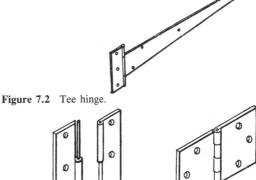

Figure 7.2 Tee hinge.

Figure 7.3 Lift-off hinge and back-flap hinge.

Parliament The parliament hinge is used to enable a door to fold back. It projects from the face of the frame (Fig. 7.4). Available in steel or brass.

Centre The centre hinge is used where it can be fixed to the top or to the side of a fitting (Fig. 7.4). Available in steel or brass.

Cranked The cranked hinge is necessary for lipped or rebated casements and is usually made with the two halves separate so that a pin fixing can be used in assembling the casement in factory production of windows (Fig. 7.5). Heavier versions are made for soundproofed rebated doors with washers and loose pins. Available in steel, brass or aluminium.

Offset The offset or *easy-clean* hinge is used to allow the outside of the window to be cleaned when open at 90 degrees (Fig. 7.5). Available in steel or aluminium.

Secret hinges Interleaving hinges are made that can be morticed into door and jamb; their origin is the USA where they are known as 'SOSS' hinges. Available in steel and bronze (see Fig. 7.46(h)).

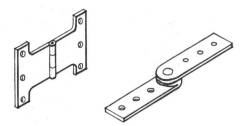

Figure 7.4   Parliament hinge and centre hinge.

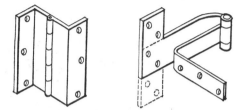

Figure 7.5   Cranked hinge and offset hinge (easy-clean).

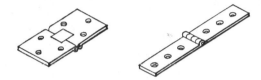

Figure 7.6   Counter-flap and strap hinge.

Counter-flap and strap   The counter-flap hinge is set in flush with the face of the work — its name being indicative of its use. The strap hinge is similarly used, but has a projecting knuckle (Fig. 7.6). Available in brass and bronze.

Friction   Figure 7.33 shows typical friction-type hinges used for pivot and tilt and turn windows. The advantage rests with the support of the sash at midspan to save the 'torque' that butt hinges would suffer under the load of wide side-hung sashes. Pivot hinges can provide complete reversibility to permit internal glazing. Pivots are friction controlled up to 5—10 degrees, safety catches limit the free opening to 100 mm. Available in steel, bronze or aluminium.

Spring   See section 7.5 (Figs 7.11 and 7.12).

7.4.2 Application of hinges

BS 1227 *Hinges* Part 1A : 1967 *Hinges for general building purposes* gives classifications according to metals as well as dimensions and weights. The Guild of Architectural

Ironmongers' publication *Hinges*, Parts 1 and 2, also provides references.

Doors and hatches   Selection of an appropriate type of hinge for a door or hatch in a particular location is determined by:

- the weight of door or hatch, including other iron-mongery such as elaborate door pulls or panic bolts
- action or other loads, e.g. floor springs and door closers
- frequency of use
- exposure to elements
- fire-resistance requirements
- burglar-resistance requirements

Typically, the weight of a 1950 mm high by 750 mm wide by 45 mm thickness door can vary between 36 and 54 kg according to whether it is half or fully glazed, or solid. When manufacturers recommend the use of a third hinge to support the weight of a door it should be positioned immediately below the top hinge (see Figs 5.18 and 7.7). However, it may sometimes be stipulated that the third hinge should be fitted centrally to stop warping in doors of slender construction or, in the case of external doors, to prevent twisting. Sherardized steel hinges provide good service internally and will resist corrosion under 'building site' conditions prior to handover and in situations like bathrooms and kitchens. Sherardized steel fittings are suitable for sheltered external locations but, ideally, steel or hinges with steel washers should not be used externally due to corrosion risks. In all cases, the size and construction of door and frame should ensure that a secure hinge fixing can be obtained.

Hinges on fire doors must have a melting point above 800°C. Suitable materials include steel, stainless steel, cast iron, phosphor-bronze and brass. The Guild of Architectural Ironmongers' publication, *Architectural Ironmongery Suitable for Use on Fire-resisting, Self-closing Timber Doorsets*, further recommends that the hinges should be at least 100 mm long, should not extend across the full thickness of the door and should leave enough room for intumescent plugs or a strip of intumescent material to be placed between the hinge leaf and the door to insulate the hinges. For practical reasons, rising butts should not be used on fire doors since door and floor springs fitted to such doors will not operate efficiently.

The need for increased security to combat the rise in crime and the need to reduce insurance premiums also mean that the choice of hinges must fulfil more than just a straight working role. The recommendations relating to appropriate quality and number of hinges on external doors needed to provide protection against burglars are covered in BS 8220 *Security of buildings against crime*, Part 1 : 1986 *Dwellings*, and Part 2 : 1987 *Offices and shops*. Figure 7.7

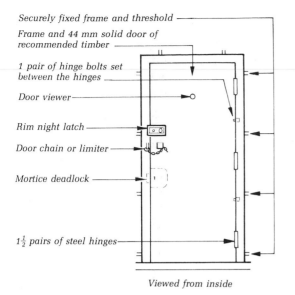

Securely fixed frame and threshold

Frame and 44 mm solid door of recommended timber

1 pair of hinge bolts set between the hinges

Door viewer

Rim night latch

Door chain or limiter

Mortice deadlock

$1\frac{1}{2}$ pairs of steel hinges

Viewed from inside

Figure 7.7 Ironmongery recommendations for security to a wood front door of a single dwelling (BS 8220 : Part 1 : 1986).

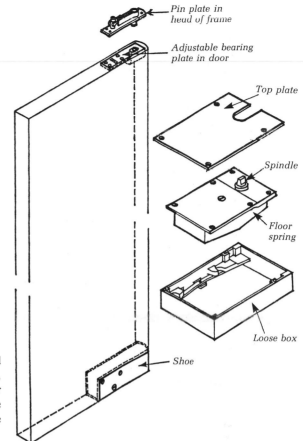

Pin plate in head of frame

Adjustable bearing plate in door

Top plate

Spindle

Floor spring

Loose box

Shoe

Figure 7.8 Typical installation of floor spring.

shows some of the recommendations relating to a wood front door of a single dwelling. As hinges can be forced, it is necessary to incorporate additional reinforcement or bolts to back up ordinary hinges and/or use hinge bolts (see Fig. 7.39). Insurance companies can insist on these provisions.

Windows Good-quality timber windows and those manufactured from metal and plastics are supplied from the factory complete with hinges and other furniture, including sophisticated friction pivot hinges, espagnolette bolting systems and locking handles (see Fig. 7.37). However, some of the cheaper ranges of wooden windows are still furnished with inadequate furniture both in terms of security and rust-free finish. It is worth seeking out suppliers who will deliver loose frames and sashes so that effective fittings can be secured on site.

7.5 Door closers and checks

7.5.1 Types of fitting

Floor springs Pivoted floor springs are the best installation for the control of swing doors since the working elements relate to the stoutest joints in door construction (namely the heavy base rail to hanging stile). They are also available with secure locks. The component consists of a strong spring contained in a metal box, a *shoe* which is

attached to the base of the door, and a top pivot. The assembly is shown in Fig. 7.8. The box is fitted into the floor thickness so that the cover plate is flush with the finished floor level; typical depths are 55—65 mm. For this reason the use of a floor spring is somewhat restricted, since many types of floor and threshold construction do not permit easy cutting away to receive the box. The adjustable pivot plate or top centre is fixed to the head of the frame and top of the door, and is adjusted up and down by a screw. The lower pivot is connected to the shoe, which is in turn firmly fixed to the bottom of the door and to the side of the door at the base. The spring should have a hydraulic check which slows down the door at a point where it still has, say, 150 mm to travel before closing. This avoids banging or injury to a person following behind.

Floor springs are illustrated in Fig. 7.9. The hydraulic check mechanism is seen in the figure as a cylinder attached by a lever arm to the strong metal springs. Double-action

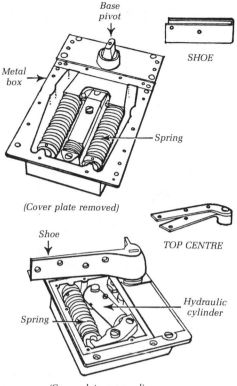

(Cover plate removed)

Figure 7.9   Double- and single-action floor spacings.

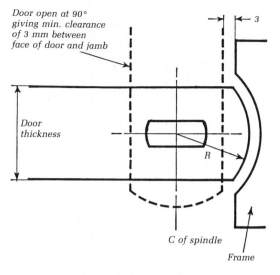

Door heels and their radii for standard applications

Door thickness	40	44	50	64
Heel radius R	32	35	38	48

Figure 7.10   Door heels and their radii.

(swinging both ways) and single-action (swinging one way only) floor springs are shown.

The cover plates which are available in a variety of finishes to match the general ironmongery specification, have been omitted for clarity. To ensure the smooth working of a double swing door in conjunction with a floor spring, it is important that both the closing and fixed edges of the door are profiled to the correct radius. The recommended dimensions are given in Fig. 7.10.

Spring hinges   There are also various types of spring hinges. Figure 7.11 illustrates a patented type of hinge controlled by a small but powerful horizontal spring held in a metal cylinder at the back of the face plate. The cylinder or cylinders have to be housed into mortices cut into the frame and are covered by the face plate. The moving part of the hinge clips round both sides of the door in a shallow housing and is screwed firmly into position so that there are no projecting knuckles or plates. Both double- and single-action hinges are illustrated, the former controlled by two springs, the latter by one. This type of hinge is not made with a check action. Another type of spring hinge

is illustrated in Fig. 7.12. This is similar in form to a butt hinge but has a large knuckle; the hinges are obtainable with double action or single action as shown. The spring, which is contained in the vertical metal cylinder, is adjustable by means of a *tommy* bar in the hole at the top of the cylinder. This adjustment controls the momentum of the closing action. Spring hinges of this type can be obtained in matching pairs, the top hinge acting as the spring, the bottom hinge being made to provide a check action. Figures 7.13 and 7.14 indicate forms of door spring closure which are recessed into the door and frame, and are independent of the hinges.

Overhead door springs   The closing action of doors hung on ordinary butt hinges can be controlled by the fixing of check mechanisms on the top of the face of the door and the door frame head. There is a very wide range of this type of overhead door closer which provides a combined closing and check action control. This type of closer is adjustable to balance the weight of the door and is much less expensive than the pivot floor spring control. The overhead closer is available in double- and single-action patterns; handed and reversible. Three alternative methods of fixing are shown. Figure 7.15 shows an example of a closer in a pleasantly designed case, for surface fixing to the opening face of the door. Figure 7.16 shows the same spring, but for fixing to the closing face of the door, and Fig. 7.17 shows a closer which fits into the thickness of

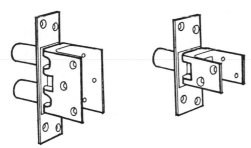

Figure 7.11 Single- and double-barrel, action spring hinges (Hawgood pattern).

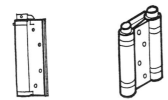

Figure 7.12 Single- and double-action spring hinges.

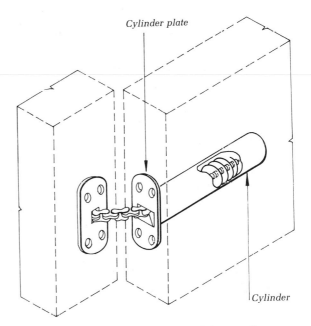

Figure 7.13 Concealed door closer for light or medium-weight internal doors (Perko pattern).

the door at the top and is thus concealed with the exception of the projection arm. The construction of the top rail needs to be strengthened to take into account the morticing, say an increase in timber depth from 95 to 175 mm. Some metal-framed doors have tubular heads designed for the inclusion of closing gear.

Door checks The door closer fittings so far described control the whole of the movement of the door, but perhaps a more universal requirement is to prevent the slamming of the door. Figure 7.18 shows such a device which, by engaging a wheel attached to a cantilevered arm, causes the movement of the door to be checked. This type of check would be used in conjunction with a spring hinge which works in conjunction with the check to achieve final and positive closing.

Where pairs of doors which have rebated meeting stiles are fitted with closing devices it is necessary to arrange that the leaves close in the correct order. To do this, a *selector* is fitted to the head of the door frame. This is a device consisting of two lever arms of unequal lengths which engage both leaves of the swing doors and can control the doors so that the rebates engage on closing. This action is shown in Fig. 7.19.

7.5.2 Application

Selection of an appropriate closer involves the following considerations:

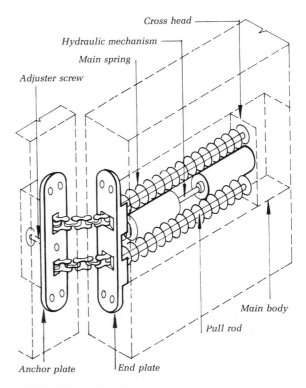

Figure 7.14 Concealed door closer incorporating combined hydraulic mechanism and spring assembly for use on fire doors (Perkomatic pattern).

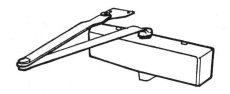

Figure 7.15 Hydraulic check, single-action door check, surface fixing to opening face.

Figure 7.16 Hydraulic check, single-action door check, surface fixing to closing face.

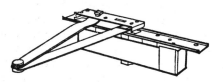

Figure 7.17 Concealed door check (fits into door thickness.)

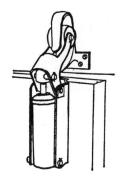

Figure 7.18 Door holder to prevent door slamming.

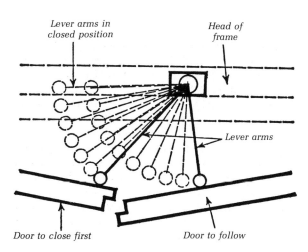

Lever arms in closed position   *Head of frame*

Lever arms

Door to close first   *Door to follow*

Figure 7.19 Door selector.

- whether exposed or concealed type required;
- ease and type of fixing possible;
- need for tamper- or vandalproof device; need for locks;
- need for adjustable device to check closing speed according to weight of door;
- incorporation of snap action which releases extra pressure during the last few degrees of closing to overcome the resistance provided by door latch;
- incorporation of a back check which brings resistance into action at a predetermined angle of opening to prevent the door opening further;
- the need for a hold-open device to allow door to remain open (not for fire doors);
- the need for a delay action which allows the door to remain open for a predetermined period before closing to allow the passage of people and goods.

Door closers must ensure the door comes to rest in a closed position. The *Building Regulations 1991* as scheduled in AD B, Appendix B requires that all fire-resisting doors should be fitted with an automatic self-closing device which

is capable of closing the door from any angle *and against any latch to the door*. However, self-closing doors usually form formidable obstacles for disabled or elderly people and they are frequently found wedged open. Under these conditions, a type of closer cap can be used which allows the door to act as a normal hinged door except when smoke or fire occurs, when a built-in detector releases a spring which closes the door. Alternatively, an electromagnetic hold-open device can be used which releases the door if a fire breaks out.

BS 5588 *Fire precautions in the design and construction of buildings* Part 3 : 1983 *Code of practice for office buildings* recommends that closers for fire doors should not be capable of being easily disconnected, should be fitted with a stand-open facility and should overcome any latches and hold the door closed in a frame until the intumescent seals have been thermally activated. The Guild of Architectural Ironmongers' publication mentioned in section 7.4.2 further recommends that concealed overhead hydraulic check-action closers should not be used for fire doors because their installation requires too much removal

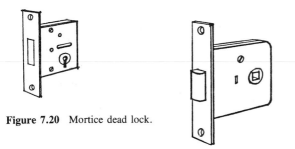

Figure 7.20 Mortice dead lock.

Figure 7.21 Upright mortice latch.

of timber in the head member of a door; springs should only be employed for FD30 doors to cupboards; and then only certain forms of floor springs should be used, i.e. those with strap shoes.

BS 6459 *Door closers* Part 1 : 1984 *Specification for mechanical performance of crank and rack and pinion overhead closers* covers mechanical back-check (this implies an inbuilt mechanism which stops the door at 90° or 100°) but not requirements specific to door closers for fire doors or other special requirements. The Guild of Architectural Ironmongers have also produced a publication on door closers and publish a Code of Practice for hardware essential to the optimum performance of fire-resisting timber door sets.

7.6 Locks and latches

7.6.1 Types of fitting

Both a lock and a latch can be supplied as separate units of ironmongery, or in combination to form a lock and latch unit. There are two methods of fixing for locks and latches in general use and the choice is whether to fix the unit on the inside face of the door or whether to set it into the thickness of the door. Where the unit is screwed to the face it is referred to as a *rim lock* or a *rim latch*; where fixed within the thickness, it is referred to as a *mortice lock* or a *mortice latch*. The projecting bolt of a rim lock or latch is retained by a *keep* fixed to the adjoining door frame, whereas it is retained by a recessed *striking plate* in the case of a mortice lock or latch. Obviously, the rim fixing is cheaper but has little security and looks a mess. On the other hand, the mortice fixing is not suitable for very thin doors: 13 mm thick locks will suit 35 mm finished thickness doors, 16 mm locks suit 40 mm doors. See also comments in sections 6.9 and 6.10.

There are five basic types from which other variations derive:

1. *Dead lock* The version illustrated in Fig. 7.20 has a single bolt which is pushed out and drawn back by operation of a key. Dead locks are so called because once the bolt has been shot it cannot be sprung back into the casing except with a key or by a *snib* or *turn* on the inside door furniture. They are generally used for securing rooms where continual free access is not a criterion. Dead bolts are sometimes designed to have *double throw* which means that the bolt goes further into the staple or keep when the key is turned a second time. This action provides added security.

2. *Latch* This is the simplest way to hold a door in the shut position. The latch illustrated in Fig. 7.21 has a bolt held in the extended position by a spring, which can be drawn back to allow the door to open by the turning of a handle or knob only. Another version is a simple 'turn bolt' or budget lock which operated by a turn knob or key; a typical application is a lavatory indicator latch.

3. *Lock and latch* The two-bolt lock illustrated in Fig. 7.22 combines the mechanisms of a dead bolt and a latch bolt. The spring latch operated by a handle serves for all general free access purposes; the dead bolt is operated by a key from one or both sides of the door for locking purposes. A dead lock suitable for a sliding door requires a claw or hook bolt to engage over an engaging device in the frame. Where pairs of doors with rebated meeting stiles are used, it is necessary to fit a rebated mortice lock. Here the fore-end of the lock case is cranked to fit the rebate on the stiles.

4. *Cylinder night latch* The *rim pattern* night latch illustrated in Fig. 7.23 has a spring bolt operated by a handle on the inside and a key on the outside of the door. When going out the door can be pulled shut behind the user, but a key is necessary for re-entering the premises. A knob or thumb slide, operated from the inside, will hold the bolt open or shut when needed and the key will then not operate. Other versions incorporating all these features are available, including a *mortice pattern* which is concealed within the door but weakens its structure, and a *narrow-stile pattern* which is suitable for use on framed doors with narrow stiles. Another version is the *dead-locking pattern* which will resist illicit activation of the latch bolt, such as by pushing it back with a knife or plastic strip inserted between frame and door, or by cutting a hole in the glass or wood panel of the door.

Figure 7.24 shows a magnetic cylinder lock. It is claimed to be practically unpickable and is more expensive than the ordinary cylinder lock. In each lock there are 14 pin positions (see later), 7 operated by one side of the key and 7 by the other. Each pin is magnetized and can only be repelled from the locked position by a magnetic force of the correct polarity, as supplied by the key.

5. *Digital and electric locks* Most lock manufacturers now produce microswitches and electric strikers for remote release. When connected with telephonic intercommunication systems, these are particularly useful for multi-storey dwellings because visitors can announce their arrival from a distance.

Time release locks are also available. Locking units that can be opened by a push-button code or a pass-card are shown in Fig. 7.25. The pass-card operation is fast becoming the normal method of control for hotel rooms, the card code can be changed by management to improve security.

7.6.2 Application of locks and latches

The following are helpful references: BS 3621 : 1980 *Thief-resistant locks* specifies design requirements, test methods and performance requirements for thief-resistant locks; BS 3827 *Glossary of terms relating to builders' hardware* Part 1 : 1964 *Locks (including locks and latches in one case)*, Part 2 : 1967 *Latches*, Part 3 : *Catches* and Part 4 : 1967 *Door, drawer, cupboard and gate furniture*, describes general concepts; BS 4951 : 1973 *Specification for builders' hardware: locks and latch furniture (doors)* gives performance tests and criteria for lever and knob furniture; and BS 8220 *Security of buildings against crime* Part 1 : 1986 *Dwellings* and Part 2 : 1987 *Offices and*

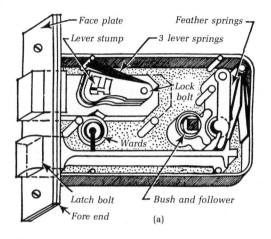

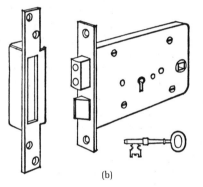

Figure 7.22 Horizontal mortice locks: (a) horizontal mortice lock showing component parts: (b) horizontal mortice lock; (c) double hook bolt for sliding doors; (d) rebated mortice lock.

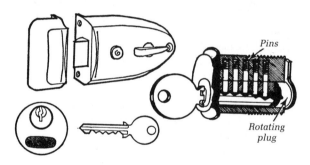

Figure 7.23 Cylinder rim latches.

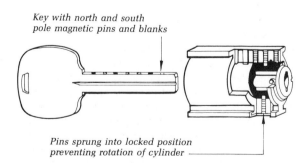

Key with north and south pole magnetic pins and blanks

Pins sprung into locked position preventing rotation of cylinder

Pins are pushed clear of shearline by like poles repelling, allowing free movement of the inner cylinder

Figure 7.24 Magnetic cylinder lock.

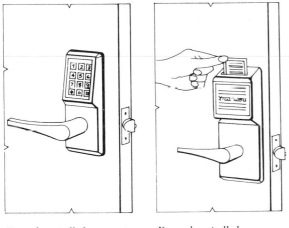

Keypad-controlled
dead-locking device

Keycard-controlled
dead-locking latch

Figure 7.25 Push-button code and pass-card locking units.

shops, contains guidance on security measures aimed at deterring burglars from entering dwellings and includes recommendations for windows and external doors. Reference can also be made to the Guild of Architectural Ironmongers' publication *Locks and Latches*, Parts 1 and 2 are the most useful.

Locking units contain complex mechanisms and great care must be exercised in the specification of appropriate quality. The specifier must be quite clear as to the precise requirements since, in general terms, cost is proportionate to security achieved. It is important to realize that, apart from the convenience of the user, the insurance company requested to cover the contents of the building will be concerned with security while the designer must be additionally concerned with providing easy means of escape in case of fire. These two conditions are not always easy to reconcile. The strongest lock fittings are made of steel with wearing parts of special bronzes.

A lock, with latch mechanism, is shown in Fig. 7.22(a). This illustrates most of the essential features. A measure of security is given by the number and complexity of the wards. If the cuts on the key bit do not correspond to the wards the key cannot be turned. The bolt is released by tumblers, or a system of levers. When the key turns, the levers have to be lifted to a certain position before the bolt will pass and so a larger number of levers gives greater security. The tumbler mechanism (pin tumblers) is applied in the normal cylinder lock as shown in Fig. 7.23(b). The V-cuts on the key have to lift the pins the exact amount so that their tops become flush with the surface of the rotatable plug to enable the latter to be turned and the latch to operate. There are many thousands of combinations of

pin positions, which gives many thousands of 'differs' or locks requiring different keys. It is important when writing the specification for the locks to be clear as to what differs are needed. In a house there is usually no point in having different room locks; in fact it is convenient, in the event of a room key being lost, to be able to use another from an adjoining room. However, on the other hand, in a building such as a hotel all room keys must differ. For this type of building locks which differ can be opened by a master key, or a number of locks, perhaps all on one floor, can be opened by a sub-master key. There is a large range of mastering 'possibles' to suit all requirements and the technique of arranging the mastering of the keys in the most convenient way is known as *suiteing*. Each group of keys being called a *suite*.

Figure 7.26 shows a typical suiteing arrangement and hierarchy of keyholders for a small factory. Some manufacturers offer a registered key system whereby the owner is registered with the merchant or supplier, and further keys are only issued to that person on receipt of his/her signature. BS 3621 describes a series of performance requirements for dead locks which, however, result in the locks not being able to have master keys.

Comments on the provision of locks and latches in connection with the need to provide effective means of escape from a building are given in section 6.8. The Guild of Architectural Ironmongers' recommendations for locks and latches incorporated on fire doors include: the maximum dimensions for upright mortice units should be 225 × 32 mm for face plate with a 160 × 19 mm case; care should be taken when fitting a mortice lock and all gaps filled with intumescent material to delay the effects of thermal bridging and fire; locks containing materials which burn or soften at fairly low temperatures (nylon and aluminium) should be avoided in fire doors because they need to be retained in a closed position during a fire; and as locking sets designed for the rebated meeting stile of double doors may affect the integrity of the door during a fire, only square-edge meeting stiles with intumescent seals should be used. Special devices used in conjunction with locks and latches help to solve the problem of allowing ease of escape for occupants while providing security against intruders from outside. Although this can be achieved by the use of *panic bolts* (see Fig. 7.38), locks can incorporate a protective glass panel on the internal face of a door which, when broken, allows access to the opening handle. Alternatively, a unit is available which allows the bolt to be automatically released when the glass is broken.

The normal mortice lock is made 'horizontal', i.e. suitable for a deep mortice into the middle rail of the door, and this type of lock is illustrated in Fig. 7.22.(a) and (b). Where a lock must fit into a narrow stile it is made 'upright'. This type of lock, which is illustrated in Fig. 7.21, has a comparatively narrow case. In a lock (lock and latch) set, the keyhole and spindle mortice are in line vertically in a vertical mortice lock, and the keyhole and spindle mortice are in line horizontally in a horizontal mortice lock. The horizontal set is usually used in conjunction with knob furniture, and the vertical set with lever handle furniture. Drawer and cupboard locks are usually for a flush fixing.

This means that they are let into the inside face of the work, so that the outside of the lock is flush with the inside face of the timber. The cover plate is usually extended round the side to give a neat finish. Figure 7.27 shows a typical cupboard lock. Simplification of the fixing is an attractive proposition as mortice cutting takes a long time. A combined lock—latch set which needs only two holes to

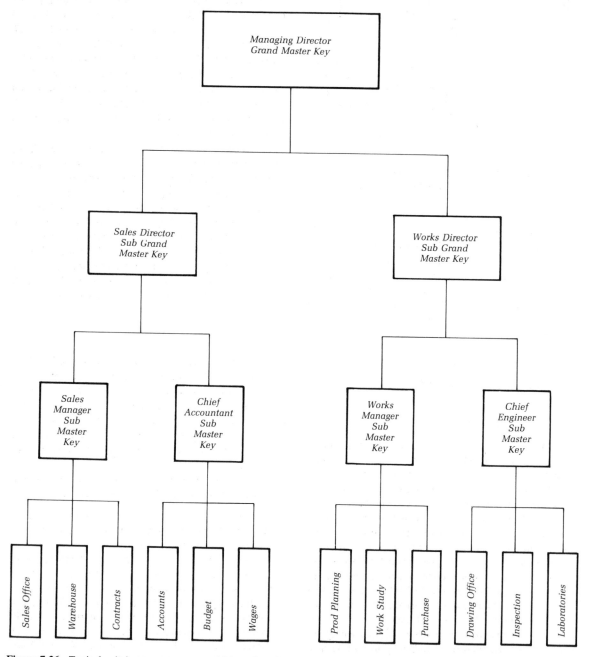

Figure 7.26   Typical suiteing arrangements and hierarchy of key holders for a small factory.

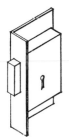

Figure 7.27 Cupboard lock.

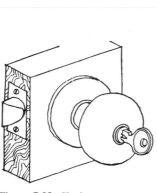

Figure 7.28   Knob set.

be drilled has been produced. This type of lock, which is now in common use for hotel bedrooms, has the locking mechanism in the knob, and is usually referred to as a *knob set*. A typical example is shown in Fig. 7.28.

Ball catches and roller catches are used for cupboards and because they are inexpensive they have also been used for the doors to living rooms in place of a latch. They are, however, very noisy and tend to give trouble in adjustment unless the projection can be easily altered to suit any change in the gap between door and frame. Magnetic catches are an alternative holding device for cupboard doors and can be fitted flush over the face of the joinery.

7.7 Knob, lever and pull handles

7.7.1 Types of fitting

The terms 'knob' and 'lever' describe the basic mechanisms for operating locks and latches. Knob and lever furniture are both available in wide ranges of shapes and finishes. It could be said that designers reveal their true level of taste when choosing hardware. Designers have revealed in times past their skill in designing door and window furniture and where ranges of patterns are still made (see Fig. 7.46). For example, the 'arts and crafts' pattern of hinges, Norfolk latches and bolts favoured by C.F.A. Voisey. By contrast, Serge Chermayeff designs for 'D' and lever handles are still manufactured. In Germany the Bauhaus ironmongery and the refinements made by Professor Burchartz are still marketed under the 'Wehag' trademark.

Round-knob furniture has the virtue that it fits any design of door, the shortcomings are difficulty in operation by young children or by the disabled. Closeness to the jamb

can damage fingers, so wide margin doors are needed to enable the correct backset of the handle to be obtained. The term 'back set' defines the dimensions from the centre of the handle spindle to the door edge. Oval knobs are easier to operate but similarly need generous backsets.

In many ways lever handles are superior in performance, but need well-sprung latches or locks to return the handle to a horizontal position. Continental latch springs are stronger than the UK designs, the mixing of handles and latches from differing suppliers not being advisable. Knob furniture need efficient springs as well, but 'tired' circular knobs do not appear unsightly. Most manufacturers offer matching ranges of ironmongery with designs and finishes to form suites of fittings. It is worth establishing how well designs have performed over the years — one should look for lever handles that fit the fingers, are shaped so that clothing is not torn and do not droop.

Two main types of pull handles are available. Those which are *D-shaped* allow a door or window to be pulled open or closed and, for doors, are often used in conjunction with push plates on the opposite side to the pull handle. The handle can be screwed to the face of a solid door, but should be bolted through glass doors or doors which are heavy. The push plate on the opposite side of the door will conceal the bolt fixing. A *cylinder pull* is located behind the external key-plate of a night latch to facilitate pulling a door to a closed position. This is necessary when a knob or lever handle is not specified for entrance doors.

7.7.2 Application

Knobs should not be used where the backset of the lock is less than, say, 60 mm. The backset is the distance from the outer face of the fore-end of the lock to the centre of the keyhole. The reason for this is if a knob set is fixed too near the door frame the user will suffer damaged knuckles when operating the knob. The question of the fixing of the knob in relation to the spindle and rose requires some special consideration.

There are many methods of fixing knob furniture, several of which were patented. The two basic variants are:

1. A spindle which is 'fixed' to the knob by a grub screw or patented fixing so that the pull of the knob is resisted directly by the spindle. This is a strong and most satisfactory method but requires exact and careful fitting.
2. 'Floating' or free spindle which slides on to the knob and which relies for its fixing by screwing the rose to the face of the door. This type is easily fitted but a disadvantage is that when used with mortice locks only short screws can be used to secure the rose because of the thickness of the lock case. These screws may work loose even in the best quality doors.

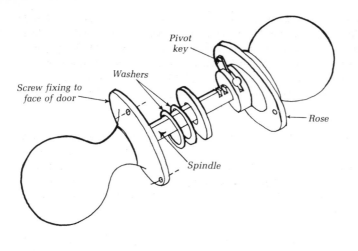

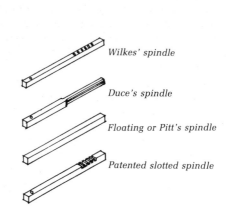

Figure 7.29 Knob furniture: aluminium handles.

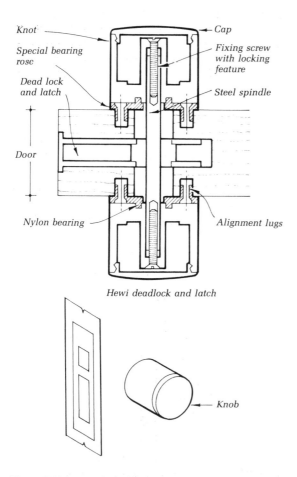

Hewi deadlock and latch

Figure 7.30 Knob furniture: plastic handles (Hewi pattern).

An 'exploded' drawing of a knob and spindle fixing is shown in Figs 7.29 and 7.30.

Knob furniture provides a neat, unobtrusive and strong specification well suited to resist rough usage. Knob furniture is available in various alloys. Bronze is the most expensive but most hard-wearing. Cheaper alternatives are iron, anodized aluminium and porcelain china, the final type being a traditional nineteenth-century form and useful in conversion work on older properties.

Lever furniture must be well designed and strongly constructed since the lever arm produces considerable strain on the lock or latch mechanism (see Fig. 7.31).

Where upright mortice locks are specified, lever furniture is essential because the distance of the spindle from the edge of the door is small. Most British lever furniture is of the floating spindle type in which the handles take the pull of the door through the handle plates or roses. Lever handles sometimes have, in place of a rose, a handle plate which allows the screws to be fixed beyond the mortice and so ensures that the fixing screws will not foul the lock case. This, of course, makes a stronger fixing. British lever handles often embody a spring to counterbalance the weight of the handle as British locks do not normally have strong springs on the latch. Continental locks have a strong latch spring and so their handles also operate on this. This point should be borne in mind when considering using continental lever handles on British locks.

In a vertical mortice lock the spindle for the lever handle or knob is vertically above the keyhole. This means that a specifier can choose between lever furniture which has a long handle plate incorporating a keyhole and a lever handle with a small handle plate and a separate key escutcheon plate. The key escutcheon plate is used as a cover plate for the keyhole. In order to hide the large number of screws which are necessary for fixing door furniture of this kind, several different types of cover plate have been produced. The cover plate will either clip over or be screwed over the fixing plates. The screw types are usually better, but can only be used where circular roses are specified. The clip-on types tend to give trouble in use unless very well designed. A drawing of a lever handle showing the fixing is given in Fig. 7.31.

Lever furniture is available in a very wide range of materials. The choice of stainless steel, bronze, nickel bronze, aluminium alloy, plastic-covered metal or nylon, will depend on considerations of first cost, appearance, type of use and subsequent maintenance costs.

7.8 Window stays and fasteners

The details for window stays and fastenings need to be read with the general review of external windows in *MBS: External Components* Chapter 4. Fittings are usually made from steel, stainless steel, bronze or aluminium due to the strength needed and the anticipated exposure.

7.8.1 Types of fitting

Stays   These are required on windows to prevent people falling out, or to prevent casements flying open in high winds. They are also used to hold the window in a limited opening position to provide ventilation. Conventional sash fasteners for top-hung windows are shown in Fig. 7.32; folding cam openers, sliding, shadbolt friction and roller stays are shown in Fig. 7.33. Of particular note is the roller stay for use on bottom-hung casements which open inwards, and on horizontal centre-pivot windows to control the projection of the top part of the window into the room.

Another form is *quadrant stays*, the traditional method used for limiting the opening of bottom-hung windows. Most stays can be supplied with an independent locking device to provide additional security to that provided by the fastener (see Fig. 7.39(d) and (g)). Some insurance companies give reduced premiums when locks are incorporated with window stays.

Fasteners   These are required to draw the casement or sash to a tightly closed position against the frame. Figure 7.34 indicates typical forms. Casement fasteners are available in wedge or cross-tongued types and some incorporate a night ventilation slot to allow enough ventilation while maintaining security.

Espagnolette bolts (see Fig.7.37(a))   These are used on high-performance windows and consist of a multi-point locking system activated from a single handle which secures the sash firmly and evenly against the seals.

Sash fasteners (see Fig. 7.39)   These are available in several types, and are fixed to the meeting rails of double-hung sash windows in order to secure both sashes in the closed positions. Forms are available which incorporate locks (see Fig. 7.39(e), (f), (h) and (i)).

7.8.2 Application

BS Code of Practice 153 *Windows and roof lights* Part 1 : 1969 *Cleaning and safety* recommends the following for tall buildings:

• Fittings for windows above third storey, and preferably above the ground floor, should limit the initial opening to 100 mm by means which cannot be tampered with by children;

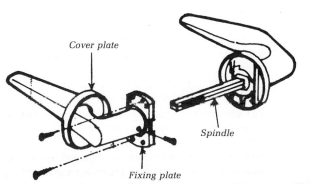

Figure 7.31 Lever furniture.

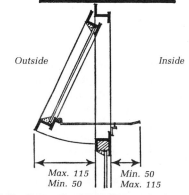

Figure 7.32 Conventional sash fastener for top-hung window.

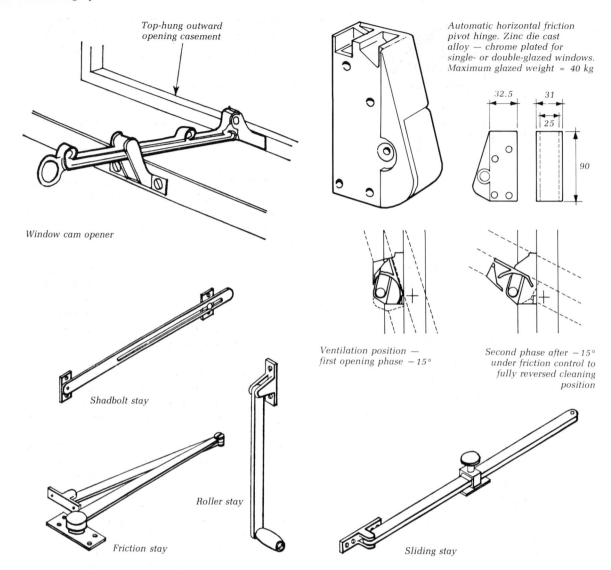

Figure 7.33 Casement and friction stays, also friction and tilt and turn hinges.

- The window should be prevented from swinging and slamming in high winds beyond the initial opening of 100 mm and friction hinges may not be adequate for this purpose;
- Large side-hung opening lights may require stays at both top and bottom to prevent distortion in high winds;
- Safety bolts or catches used for locking pivoted windows in their reversed position should be positive in action and not liable to give way under pressure. The lives of window cleaners often depend upon such catches working properly.

7.9 Bolts

7.9.1 Types of fitting

Bolts form additional locking devices which can only be operated from the inside of a building.

Barrel This is the most common type and has a round or barrel-like shoot on the back-plate for surface fixing, as shown in Fig. 7.35. The shoot runs in a guide and is slid home into a metal keep. This type of bolt is inexpensive

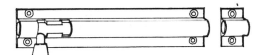

Figure 7.35 Barrel bolt.

and easy to fix. The size and pattern are often laid down by insurers.

Flush lever   This is shown in Fig. 7.36. It is recessed into a shallow housing in the component to be secured until the face plate is flush with the surface of the timber. The bolt is operated by a thumb slide or lever action. Flush lever bolts are more expensive than barrel bolts and take more time to fix.

Cremorne   This is a particular type of surface espagnolette bolt used for minimizing the twisting of a door or window (see Fig. 7.37(b)). It extends the full height of the door or window so that when the handle is turned the top bolt slides upwards and the bottom bolt slides downwards to give top and bottom fixing.

Espagnolette   This type of bolt provides centre fixing as well as fixing at the top and bottom and to the jambs if needed. The centre fixing is commonly also a lock. Espagnolettes may be surface or flush fitting as required.

Panic   Ordinary bolts must not be fixed on doors which are used as means of escape in case of fire. To overcome the problem of security, a *panic latch* or a *panic bolt* is employed (see Fig. 7.38). Panic latches are used on single doors and consist of a crossbar which is pushed against a latch to release it. A locking knob is often fixed to the outside of the door to enable two-way traffic to operate. Panic bolts are used on single and double doors, and have a striking plate at the top and bottom so that the door is held in three places, thus giving a greater degree of security. A mortice panic bolt is let into the face of the door for neatness. Break-glass fittings also exist where escape doors are only intended for use in emergencies.

Security   This is the name given to those bolts which are usually morticed into the construction of a door or window or are fixed to their stays and fasteners in order to provide greater security. Figure 7.39 shows a typical range.

7.9.2 Application of security bolts

The principles involved in the application of bolts to

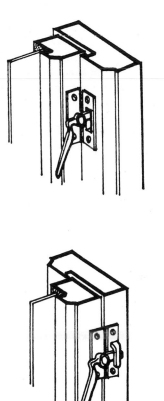

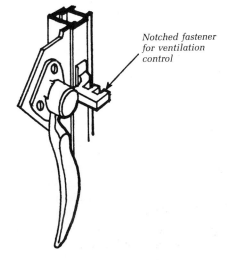

Notched fastener
for ventilation
control

Figure 7.34 Casement turns or fasteners.

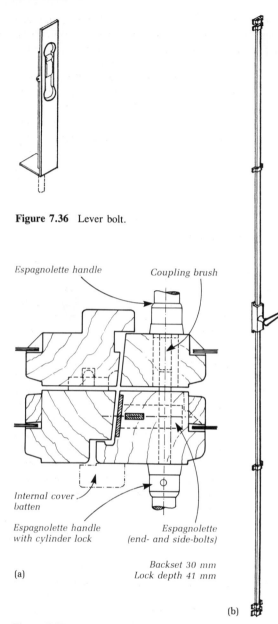

Figure 7.36 Lever bolt.

Espagnolette handle

Coupling brush

Internal cover batten

Espagnolette handle with cylinder lock

Espagnolette (end- and side-bolts)

Backset 30 mm
Lock depth 41 mm

(a)

(b)

Figure 7.37 (a) Cremorne bolt; (b) morticed espagnolette bolt in meeting stile of double-glazed french windows.

windows and doors are mostly self-evident. They are usually fixed at the top and bottom of doors and should always have a socket to receive the shoot of the bolt. This is particularly important at the threshold. The diameter of the shoot, the type of metal used and the method of fixing of the bolt are an indication of its strength. For double doors it is normally necessary to secure the first closing leaf with bolts to enable the doors to be locked.

There are various designs of panic latches and panic bolts to suit different degrees of security. The height of the push-bar is important since in an emergency it must operate when people fall against it — the generally accepted height being 1050 mm above floor level. BS 5725 *Emergency exit devices* Part 1 : 1981 *Specification for panic bolts and panic latches mechanically operated by horizontal push-bar* gives maximum acceptable forces needed to operate them as well as specifying the strength and durability of components, heat resistance of materials and methods for fitting the equipment. Panic bolts can be difficult to operate unless regularly cleaned and maintained, particularly the lower striking plates. Panic latches can have a locking device on the external face for routine access.

7.10 Miscellaneous items

7.10.1 Cupboard catches and devices

Manufacturers' catalogues should be studied to understand the wide selection. It is important to establish the principle to be adopted — the combined handle and catch, usually of the press knob variation and centrally mounted on the closing side of the door. Lay-on spring hinges have largely replaced catches and will hold the door shut. They will also give greater freedom of choice of handle, knob, recessed grip or 'D' pattern. Magnetic catches can form a useful reinforcement to spring hinges, but should be mounted in pairs to prevent warping of the doors. Some forms of catch are activated by pressure on the door front which removes the need for any external furniture (see Fig. 7.40). Telescopic guides exist for drawers and for space-saving devices. Interlocking telescopic guides enable an interleaved table top to be packed into a 600 mm space within cupboard framing.

7.10.2 Sliding gear

Sliding gear for doors and windows has been described separately with the fittings illustrated in the appropriate chapter. Sliding gear for cupboard doors is available in a very wide range. Small cupboard or bookcase fronts of plate glass with polished edges can be fitted directly into the channels of fibre, metal or plastic made in single, double or triple section. Thin aluminium or plastic-faced sheet, or plywood, can also run in most of these tracks. Typical sections of this type of track are shown in Fig. 7.41. For larger plate-glass doors a metal section track is provided in aluminium or brass which incorporates small wheels or ball-bearings that run on a bottom track to take the extra weight of glass.

For larger plywood or blockwood, or framed cupboard

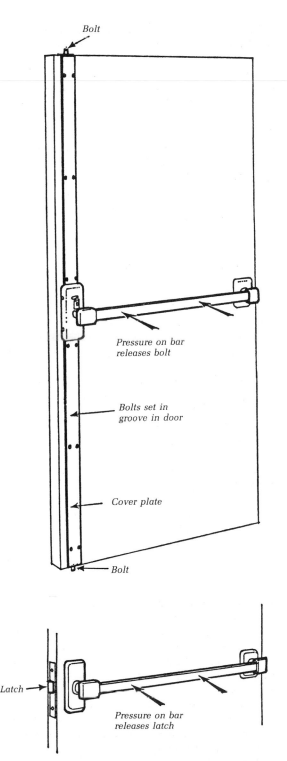

Figure 7.38   Mortice panic bolt.

doors, a fibre track with sliders is manufactured. The track is grooved into the sill, the sliders being morticed into the under edge of the door. This type of track is also made in nylon and an example is illustrated in Fig. 7.42. As an alternative to this there are a number of small ball-bearing roller fittings, for running on a bottom track, which are illustrated in Fig. 7.43. These run easily and so are used where the door is tall in proportion to its width, so might jam in a simple channel track. Cupboard doors that are 2.1–2.4 m high should be treated as sliding room doors and be suspended from an overhead track with bottom guides; sliding folding doors can also be operated with such tracks.

7.10.3 Window-opening gear

The rod and worm gear type of control has been traditional for large and heavy windows — it is suitable where cost is the main consideration and a neat appearance is not essential. Regular maintenance must be organized since the working parts must be kept clean and well oiled. If they seize up the fixings will be wrenched from the wall by forcing the gearing. Alternative systems in common use comprise a special wire cable sliding in a metal tube. A system in which the cable is wired to serve efficiently both in compression and in tension is shown in Fig. 7.44. The wire operates directly on the window and is in turn worked either by a slide for small installations or, in the case of heavier windows, by a geared regulator. There is a limit to the range of windows which can be controlled by mechanical means, and for very large installations travelling over long distances electrical or hydraulic systems must be used. This type of system is, however, uneconomical for small installations.

For this type of control a motor is installed at the receiving end which will drive a local installation of cable gear. The main push-button control can be situated in a convenient central position and is coupled to a forward and reverse contactor. The current is switched off in both directions by micro-switches at the receiving end.

Hydraulic control is achieved by a small-bore copper nylon tubing filled with oil. A pump, either hand or electrically operated, delivers the requisite pressure to small hydraulic rams positioned in the actual opening gear. In this system a single operating position can be used to control a large number of opening lights remotely situated both from the operating position and from each other. Because of their neat appearance both electric and hydraulic systems are preferable provided that their initial cost can be justified.

7.10.4 Door viewers

These are small telescope devices fitted through solid doors

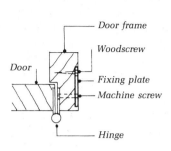

(a)

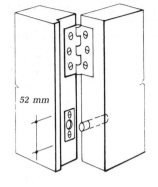

(b)

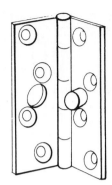

(c)

DOORS

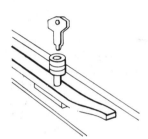

(d)

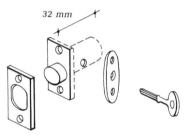

(e)

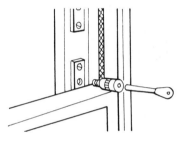

(f)

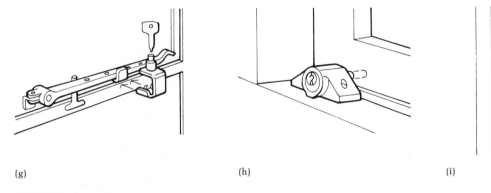

(g)

(h)

(i)

WINDOWS

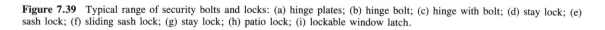

Figure 7.39 Typical range of security bolts and locks: (a) hinge plates; (b) hinge bolt; (c) hinge with bolt; (d) stay lock; (e) sash lock; (f) sliding sash lock; (g) stay lock; (h) patio lock; (i) lockable window latch.

to allow the occupier to identify visitors without opening the door.

7.10.5 Door chains

These chains are fitted to a door to prevent unauthorized forced entry when the visitor cannot be seen without opening the door. Their success depends on the adequacy of the screw-fixing plates and the quality of the metal used for the chain.

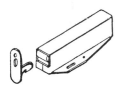

Figure 7.40 Cupboard catch.

7.11 Scheduling ironmongery

The usual method for specifying ironmongery involves the production of a schedule which locates the door or window by number, and then by type, fire resistance and frame details before listing the required ironmongery or code for an ironmongery package (Fig. 7.45). Alternatively, individual data sheets are produced for each door or window. These schedules are now frequently produced with the aid of computer programs.

Many fittings, such as locks and handles, are handed. This means that they are specifically for a door hung either on the left side or the right side of an opening. Therefore, it is essential to have a standard way of describing on which side the door is hung. It is usual to describe the direction of opening as *clockwise* or *anticlockwise* when viewed in plan from the outward opening position. The clockwise part

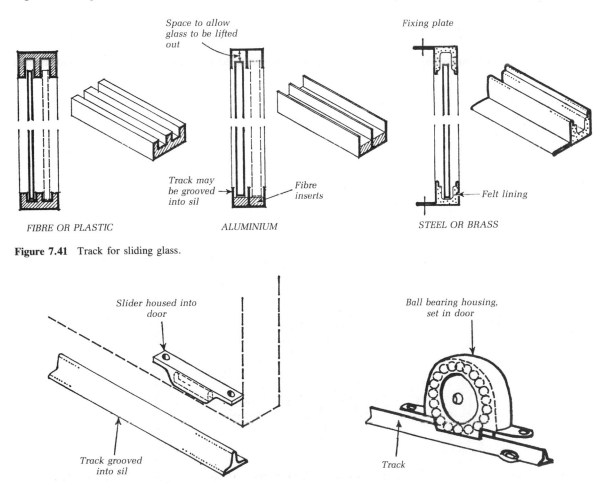

Figure 7.41 Track for sliding glass.

Figure 7.42 Door sliders and track.

Figure 7.43 Cupboard ball-bearing roller track.

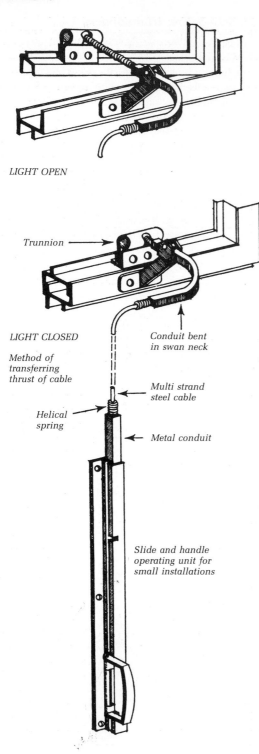

LIGHT OPEN

Trunnion ⟶

LIGHT CLOSED

Conduit bent
in swan neck

Method of
transferring
thrust of cable

Multi strand
steel cable

Helical
spring

Metal conduit

Slide and handle
operating unit for
small installations

Figure 7.44   Remote control device for high-level opening
windows.

is self-evident, but the 'outside' is generally agreed in plan
as follows:

- for external doors — the 'open-air' side;
- for internal doors — the corridor side of a room;
- for cupboards — the room side.

When describing locks, the door is viewed from the outside
and a lock on the left side will be a left-hand lock.

7.12 Designer's skill

There is finally the impact of the designer's skill on the
appearance of ironmongery fittings. The 'arts and crafts'
tradition was maintained by Sir Edwin Lutyens who drew
upon handmade wrought ironwork for hinges, latches and
window fasteners, with often each door or window fitted
with individually designed fittings (Fig. 7.46(a)). Other
architects like Adams, Holden and Pearson (associated with
Frank Pick of the London Passenger Transport Board in
the 1920s) inspired manufacturers to produce ranges of
handles and door plates which carried the 'hallmark' of the
designer. In the UK, makers such as J. D. Beardmore are
still marketing 'Norfolk' latches that owe their origins to
the studio of C. F. A. Voisey. The Bauhaus originated
simple tubular designs, which employed a 'rod' handle,
with an oversleeve, today often plastic or timber, Figure
7.46(d) and (e) demonstrates variations on this theme
available from Sweden.

Many patterns were developed for particular commissions
and the office of Mendlesohn and Chermayeff were
responsible for new suites of furniture made by Yannedis
& Co. Ltd in the 1930s. Figure 7.46(f) shows 'D' handles
for cupboard fittings, dating back to 1935. A leading
German firm invited architects to participate in design
development, some of the classic 'Wehag' range of door
and window furniture dating back to the early 1930s. The
most distinguished work still available is credited to
Professor Burchartz as featured in Fig. 7.46(g). British
ironmongery firms have followed this lead, the most notable
being W. & S. Allgood with the Modric ranges designed
by Alan Tye Design RDI Ltd since the late 1970s.

Finally, there is the inventive and innovative aspect and
this is demonstrated by the 'secret' folding interleaf device
known as the 'SOSS' hinge, patented by Arnold & Co. in
the USA many years ago.

Figure 7.45   Ironmongery schedule.

DOOR GROUP PREFIX LETTERS			INTERNAL DOORS													EXTERNAL DOORS coded XD												
DOOR NUMBERS			1.	2	3	4	5	6	7	8	9	10	11	12	13	1	2	3	4	5	6	7	8	9	10	11	12	13
COMMON SUITING MASTER KEYING.	REFERENCE NUMBERS	AZA								DOUBLE						DOUBLE				DOUBLE		DOUBLE		DOUBLE.				
UPRIGHT LOCKS	WITH ONE KEY	100																										
	LOCKS TO PASS	101	1	1	1		1					1	1	1														
	WITH TWO KEYS	102																										
	REBATED COMPONENTS	103																										
	ROLLER BOLT, ONE KEY	104																										
	ROLLER BOLT, LOCKS TO PASS	105																										
	ROLLER BOLT, TWO KEYS	106																										
	ROLLER BOLT REBATED COMP'S	107																										
UPRIGHT DEADLOCK	WITH ONE KEY	112																										
	LOCKS TO PASS	113								1.						1.		1	1	1		1		1		1	1	1
	WITH TWO KEYS	114																										
	REBATED COMPONENTS	115														1		1	1	1		1		1		1	1	1
LEVER HANDLES	PAIR ON ROSE	130																										
	PAIR ON BACKPLATE, KEYHOLE	131	1	1	1		1					1	1	1														
	PAIR ON BACKPLATE, NO KEYHOLE	132																										
	ESCUTCHEONS	133								2						2		2	2	2		2		2		2	2	2
PULL HANDLES	150 mm CENTRES FIXING	134																										
	225 mm CENTRES	135						1	1	1	1				1	1	1	1	1	1	1	1	1	1	1	1	1	1
	300 mm CENTRES	136																										
FINGER PLATES	300 x 75	140				1		1	1	1	1				1	1	1	1	1	1	1	1	1	1	1	1	1	1
KICKING PLATES	625 mm WIDE	141													1.													
	725 mm	142				1	1	1	1	2	2		1															
	775 mm	143	1	1	2							2		2						1	1					1	1	
	825 mm	144														1	1	1	1			1	1	1	1			1
	875 mm.	145																										
FLUSH AND BARREL BOLTS	PAIR FLUSH BOLTS	150								1						1				1		1		1				
	SOCKET FOR WOOD	151																										
	SOCKET FOR CONCRETE	152																										
	PAIR BARREL BOLTS	153																										
OVERHEAD CLOSERS	FOR EXTERNAL DOORS OPEN OUT	160																										
	FOR INTERNAL DOORS	161				1		1	1					1	1													
	DOOR SELECTOR	162																										
	OVERHEAD LIMITING STAY	163														1	1	1	1	1	1	1	1	1	1	1	1	
DOOR STOPS	FOR TIMBER	170																										
	FOR CONCRETE	171				1						1	1	1	1													
	POST MOUNTED DOOR HOLDER	172																										
	CABIN HOOK	173																										
LETTER BOX	LETTER BOX.	174																									1	
HAT AND COAT HOOKS	ALUMINIUM	180																										
	NYLON COATED SECRET FIX.	181																										
	PAIR NYLON COATED SECRET FIX	182																										
	NYLON COATED SCREW FIX.	183																										

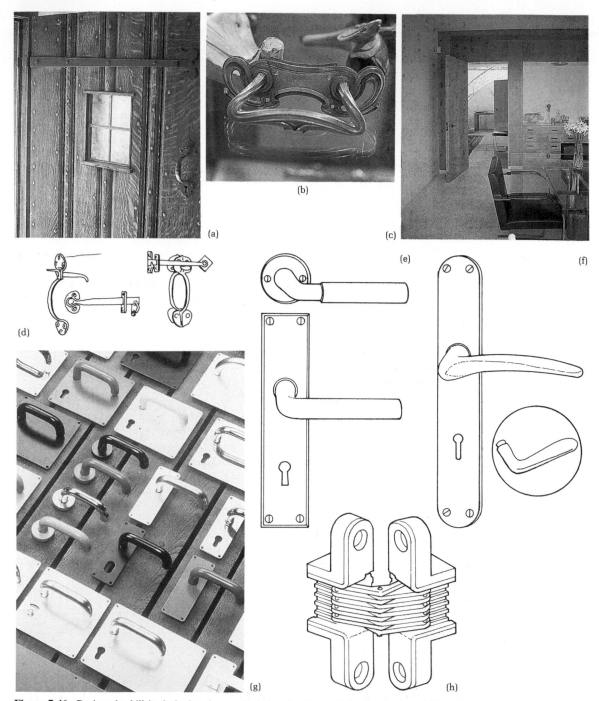

Figure 7.46 Designer's skill in designing door and window furniture: (a) handmade wrought iron door fittings that differ from door to door (Goddards, 1899; architect: Sir Edwin Lutyens); (b) bronze ironmongery designed by Victor Horta (*c.* 1890s); (c) 'D' handles, house at Chalfont St Giles (1934) (architects: Mendlesohn and Chermayeff; maker Yannedis & Co. Ltd); (d) Norfolk latches of wrought iron in the style of C. F. A. Voisey (*c.* 1890s) (today made in cast iron); (e) Bauhaus designs (*c.* 1924) made today in Sweden (materials nickel bronze; maker GKN-Stenman AB); (f) 'Wehag' 70L range from the early 1930s designed by Prof. Burchartz (maker Wilh. Engstfeld GmbH & Co.; distributed by Laidlaw and Thomson Group; material silver anodized aluminium) (g) Modric range designed by Alan Tye Design RDI Ltd for G. & S. Allgood Ltd in a wide range of finishes; (h) 'SOSS' interleaf hinge patented by Arnold & Co. USA (material bronze or stainless steel).

8 Stairs and balustrades

8.1 Introduction

Staircases can be subdivided into two forms of construction, those integral with the building structure and those which are self-supporting, either propped against walls (like ladders) or between floors or framed independently. External and landscape stairs have similar categories but are excluded from this chapter apart from fire escape components. A schedule of common terms used in stair construction is given in the following summary.

Apron lining Lining used to finish exposed edge to floors at stair-well.

Apron panels The infilling to the triangular area between strings and the floor line.

Balustrade Guarding employed at open sides to stairs and landing.

Brackets Timber supports for treads that are secured to carriage pieces.

Carriage pieces Timber beams or bearers inserted below treads and risers to improve support at mid-spars in wide timber stairs.

Flight A general description for a sequence of steps between landings.

Going The horizontal dimension from nosing edge to nosing edge.

Handrail The top member of a balustrade or wall rail to provide a handhold.

Helical stairs Cylindrical stairs which are constructed within a constant drum.

Lining An abbreviated term for apron lining.

Newel The post employed at the base and head of stair balustrades to help brace the balustrade.

Nosing The extreme leading edge of a step or landing.

Nosing line The setting-out line for the landing edge of nosings.

Part riser Used to reduce the gap between treads in open stair construction.

Pitch The angle contained by the nosing edges and the horizontal floor plane.

Shaft The total space occupied by stairs and landings.

Spirals Stairs constructed within a tapering drum. The term is commonly applied to free-standing stairs supported off a central post and employing a constant geometry.

Standard A term employed for metal supports to balustrades, where length of construction depends upon larger members for bracing.

String The framing members at the edge of flights used to carry the treads and risers. Can be formed in timber, metal or reinforced concrete.

Tapering treads See windows.

Three- and four-turn stairs Stairs which change direction three or four times around a central well.

Tread The horizontal dimension from nosing edge to nosing edge; also applies to the horizontal face or component of each tread.

Well A general term that can apply to the whole space or shaft occupied by stairs, or to the clear space between stairs and open landings or between stairs in dog-leg flights.

Winders Tapering treads which occur at landings or at the base of stairs.

Wreathing Curved handrailing in metal or timber where handrails curve in one or more directions at the base and head of stair balustrades.

The integral structure of building and stairs occurs historically with turret stairs, the spiral form favoured by builders of church towers and castles. Stone treads are built as slabs spanning between a central pier and the enclosing drum of masonry (Fig. 8.1(a)). Another version, known as helical stairs, employs cantilever slabs projecting from the drum walls with an open well at the centre (Fig. 8.1(b)). A similar technique is seen in London houses of the eighteenth and nineteenth centuries where cantilever stone stairs are found in the flights from basement, first or second floor. Brick and stone masonry vaulting played a significant role in stair construction in times past; a typical pattern from the Renaissance occurred with sloping vaults springing from arches and piers that enabled a multi-storey flight of stairs to be constructed as a complete *treppenhaus* (meaning stair hall), illustrated in Fig. 8.1(c).

Today stairs and lift shafts which form structural cores are often constructed with *in situ* reinforced concrete framing to provide structural concrete sheer walls. The interaction of the vertical structural walls enclosing stair and lift shafts against the floor framing provides stiffness to the total frame as shown in Fig. 8.1(d). Casting reinforced concrete stairs between the shaft walls increases

rigidity. Cores of this pattern in reinforced concrete are often employed compositely at the centre of steel-framed buildings.

Stair components such as precast concrete flights or folded steel plate stairs are often industrialized and brought to site to be propped between erected frames (Fig. 8.1(e)). Timber stairs are also made away from site for propping in place.

Self-sufficient structures exist such as spiral stairs which depend upon a central column of reinforced concrete, steel or a timber drum to support the cantilever treads. Structures which rely upon reinforced concrete and steel of this form can rise 12–15 m (Fig. 8.1(f) and (g)). More conventional orthogonal framing can be arranged with tubular steel columns or rolled steel joists to support dog-legged stairs

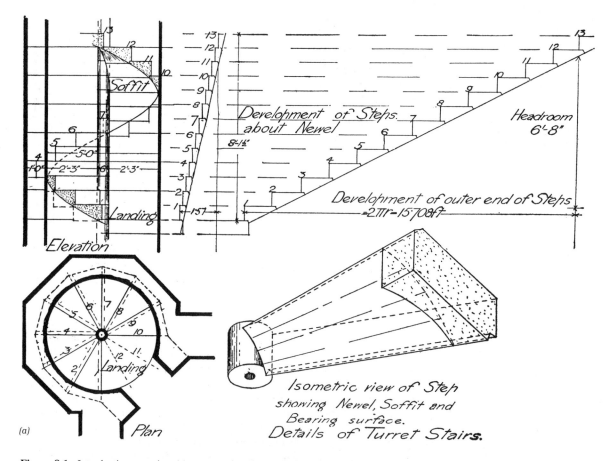

Figure 8.1 Introduction to stairs: (a) turret stairs (from Mitchell's *Building Construction*, vol. 2, 1944 edn); (b) open well cantilever stairs, Queen's House, Greenwich, London (architect: Inigo Jones); (c) *Treppenhaus*, stair structure with vaults and arches; (d) stiffening role of reinforced concrete core for lift and stair shafts; (e) precast concrete industrialized flights; (f) precast spiral stairs and central column (photo: Alistair Chrew-Cox); (g) steel-framed fire escape stairs.

(c)

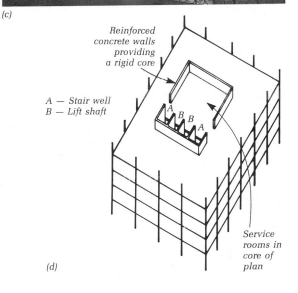

Reinforced
concrete walls
providing
a rigid core

A — Stair well
B — Lift shaft

Service
rooms in
core of
plan

Figure 8.1 *continued*   (d)

favoured for external fire escape stairs. Such steel framing
can also be used for internal stairs in conjunction with the
general framing of the structure.

8.2 Performance requirements

8.2.1 Building Regulations (see Fig. 8.2(a)–(f))

The Building Regulations categorize stairs according to their
use in buildings. Essentially, private domestic versus public
use, and then subdivisions determined by specific
occupancy and utilization of spaces served by the stair in
question. The performance standards in terms of tread and
rise proportions, pitch, width, headroom and height of
guard rails are set down in Part K of the 1991 Building
Regulations. The principal diagrams are illustrated in Fig.
8.2(a)–(d).

Ladders, minimal spiral stairs of 1.4 m diameter and
narrow alternating tread flights are permitted for
maintenance or for occasional access to minor
accommodation. The designer needs to look at the permitted
relaxations under Approved Document K1, sections 1.10,
1.21, 1.24, 1.25 and 1.26.

Multipurpose buildings with stairs connecting various
categories of use will have to adopt a common denominator
in stair risers. This means that standardized stairs can be
made and a modular approach adopted to vertical
dimensions within the total building section. Under the
Building Regulations, the 180 mm riser × 280 mm tread
provides the statutory minimum stair proportion for all
categories of use within a multipurpose design (e.g.
domestic, assembly and institutional use and all other uses).
There are many circumstances where this proportion of rise
to tread may be considered too abrupt and uncomfortable
as compared with the 150 mm × 300 mm proportion used
by Wren.

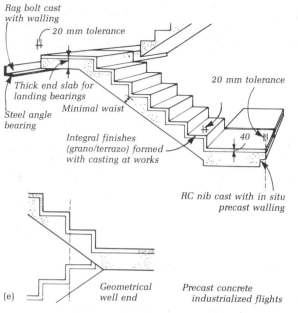

Rag bolt cast
with walling

20 mm tolerance

Thick end slab for
landing bearings

Steel angle
bearing

Minimal waist

20 mm tolerance

40

Integral finishes
(grano/terrazo) formed
with casting at works

RC nib cast with in situ
precast walling

(e)

Geometrical
well end

Precast concrete
industrialized flights

(f)

(g)

Figure 8.1 *continued*

The mandatory geometry affects the numbers of risers between landings and the permitted tapers in spiral and winding stairs, as illustrated in Fig. 8.2(e) and (f). The requirements for guarding of stairs and landings are detailed later in this chapter (section 8.8). One of the more onerous safety aspects concerns the maximum gap between stair components to prevent the trapping of children. The regulations call for a clear space that will not pass a 100 mm diameter sphere. This aspect affects the design of open tread stairs (see Fig. 8.7(a)), and of course balustrades wherever children may be at risk. See section 8.8 for details.

8.2.2 Staircases and means of escape

These complex provisions cannot be summarized in a few paragraphs — the designer has to look at the Regulations in detail. The most useful guide which covers the legislation up to December 1992 is contained in a publication *Guidance Document to the 1991 Building Regulations*.[1] The situation is complicated by the differing interpretations drawn up by fire officers, the way forward in the design process being

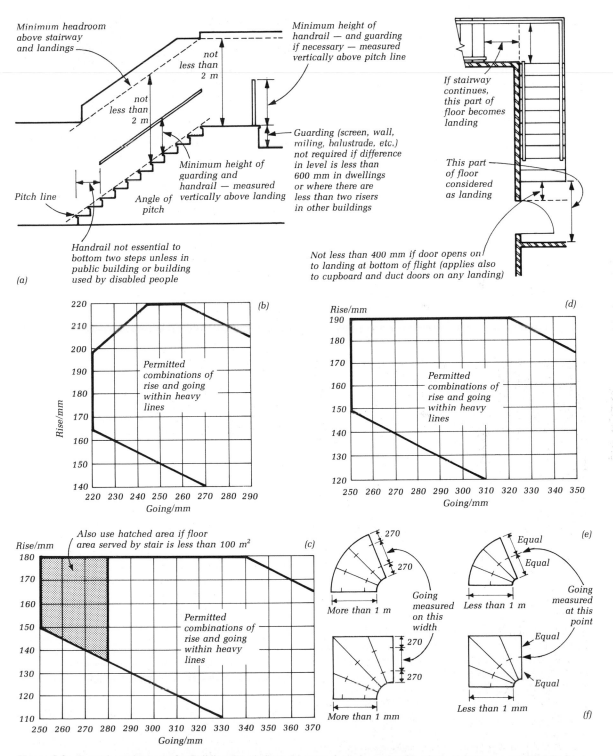

Figure 8.2 General requirements for Building Regulations: (a) general requirements; (b) private stairways — practical limits for rise and going; (c) stairs in institutional and assembly buildings — practical limits for rise and going; (d) other buildings (except (b) and (c)) — practical limits for rise and going; (e) tapered treads of equal width; (f) tapered treads of unequal width.

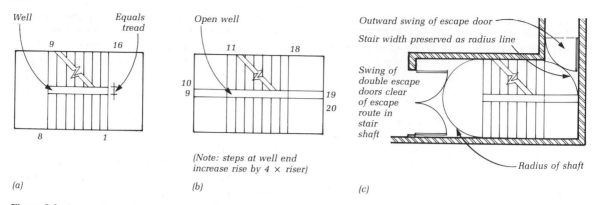

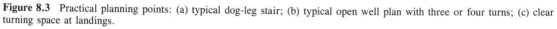

Figure 8.3 Practical planning points: (a) typical dog-leg stair; (b) typical open well plan with three or four turns; (c) clear turning space at landings.

negotiation until an agreed layout can be settled. The details will need to schedule numbers of steps, tread to rise relation, pitch, effective widths, headrooms, guarding arrangements, materials and finishes, fire resistance to stair enclosures and forms of fire doors, exit routes and emergency lighting. The accompanying plans will need to illustrate the total floor plans and sections since the extent and volume of the building dictate the girth and number of escape stairs.

Codes of Practice call for fire-resisting escape stairs or for limited combustibility, but are not specific as to the required construction. It is customary to use reinforced or precast concrete with steel balustrading, or else steel framing with welded plate treads and risers. The structural steel will be protected by intumescent paint while treads and risers will have non-combustible cladding of tile, etc. Hardwoods of 45 mm minimum thickness are permitted for overlaying treads and for handrails. The use of aluminium is a matter of negotiation, due to loss of strength under fire tests.

Practical points concerning the satisfactory arrangement of an escape stair are listed below. This should be seen as a starting point in the design process and as an aid to disentangle the confused regulations. Figure 8.3(a)−(c) provide the illustrations.

- Logical layouts must be developed which permit the escape route to follow a clockwise or anticlockwise direction, ideally without contra-movement to the final exit. Remember that three- and four-turn stairs can cope with varying heights floor to floor within a constant shaft size (Fig. 8.3(a) and (b)). Three- and four-turn stairs imply the use of steps within the landing spaces at either end.
- Maintain constant tread to rise relationship to whole escape stair, avoiding single steps. Remember that two risers is the minimum flight permitted. Use ramps at 1:12 slope where single steps might otherwise occur.

Ensure that minimum headroom is not infringed by beams. Avoid tapered and spiral stairs for escape stairs.

- Maintain stair widths (measured centre of handrail to centre of handrail) and increase as required by traffic. Also maintain widths of landings and allow clear turning circle without obstruction from door swings. Allow all doors to open in the path of escape (Fig. 8.3(c)).
- In buildings with a single escape stair allow for a suitable window at each landing for entry by fire-fighters.
- Provide handrails that match the length of each flight and are designed to give handhold without obstruction from the surrounding structure. Ideally the handrail should run smoothly at well ends without awkward steps.
- Consider arrangements for natural lighting, artificial lighting and emergency illumination. Top lighting to an open well stair is often very effective.

8.2.3 British Standards and Codes of Practice

It is helpful to understand the standard definitions before delving into British Standards and Codes of Practice. A full reference is given in BS 5578 : Part 1 : 1978, and an introduction in section 8.1.

Private stairs Stairs in or serving only one dwelling.
Institutional and assembly stairs Stairs serving places where a substantial number of people gather.
Other stairs Stairs serving all other buildings apart from those referred to above.
Flight Section of a ramp or stair running between landings with a continuous slope or series of steps.
Going The distance measured in plan across the tread less any overlap with the next tread.
Rise The vertical distance between the top surfaces of two consecutive treads.
Pitch line May be defined as a notional line connecting

the nosings of all treads, including the nosings of the landing at the top of the flight.

Tapered tread A tread with a nosing which is not parallel to the nosing of the tread or landing above it.

Reverting to British Standards the following are useful in designing stairs to accord with building categories:

- BS 5588 : Section 1.1, 1984 Code of Practice relates to stairs for dwelling-houses of three or more storeys.
- BS Code of Practice CP 3 : Chapter IV : Part 1 : 1971 relates to stairs for a flat or flats in three or more storeys.
- BS 5588 : Part 2 : 1985 relates to stairs for one shop or shops.
- BS 5588 : Part 3 : 1983 relates to stairs from one office or offices.
- BS 5588 : Part 5 : 1986 relates to fire-fighting stairs and lifts in residential buildings.
- BS 5588 : Part 6 : 1991 relates to stairs for places of assembly.
- BS 5588 : Part 8 : 1985 relates to means of escape for the disabled.
- BS 5810 : 1985 relates to access for disabled to buildings.
- BS 5619 : 1978 Codes of Practice for design of housing for the convenience of disabled people.

Other British Standards relate to specific details:

- Ladders, walkways, industrial-type stairs — BS 4211 : 1987 and BS 5395 : Part 3 : 1985.
- Wood stairs — BS 585 : Part 1 : 1989 and Part 2 : 1985.
- Helical and spiral stairs — BS 5395 : Part 2 : 1984.
- Modular co-ordination — BS 5578 : Part 2 : 1978.
- Protective barriers in and about buildings (these relate for example to landing and stair balustrades) — BS 6180 : 1982. States required strengths for balustrades.

8.3 Methods of constructional design

Valuable references exist in eighteenth and nineteenth century construction books to traditional assemblies for staircases, and particularly to the complex setting out needed for curving or tapered stairs and their related handrailing and strings. Cost of printing and copyright prevents wholesale reproduction of this source material, but specialist manufacturers have incorporated the key information into computer-aided design programs. Further reading on source material is given at the end of the chapter.[2]

Computer-aided resources from the industry include feasibility studies to establish stair geometry which fulfils the ratios and pitch limitations of the Building Regulations; these restrictions assume significant proportions in laying out tapered steps or spiral designs. A typical computer-generated layout is given in Fig. 8.4(a) for spiral stairs. Visual assessment of details is best arranged through full-size drawings for items such as step profiles, balustrade and handrail pattern, while a full-size mock-up in real or simulated materials may be needed for judgement to be made in important locations. Computer-generated drawings can explore the geometrical modelling from all directions, for example, stair soffit to string and landing relationship. Another significant area that can be explored is the balustrade pattern and handrailing in relation to nosing line and to the problems created at landings where the design rhythm needs to be maintained.

There are simple devices in layout which facilitate a simpler geometry for handrailing where dog-leg stairs are planned. The choices are illustrated in Fig. 8.4 (b) and (c). The first layout is the most spacious. It involves a well depth equal to the tread dimension so that the rising line of the handrail at the well end is sympathetic to the pitch of the dog-leg flight, thus creating a smooth transition for the balustrade pattern from floor to floor. This plan does, however, add 250–300 mm to the effective length of the stair shaft due to the extended well ends. The second variation involves a half-step relation across the well, while the well end handrail is run horizontal to complete the alignment. In this case the half step stagger only adds 125–150 mm to the shaft.

The use of newel posts at landing ends can accommodate the changes of balustrade height that arise between flight and landing under the current Building Regulations but which provide an awkward break in the handholds. Figure 8.4(d) and (e) explores three traditional ways of solving handrail design in tight situations. Providing a solid balustrade wall or screen independent of the handrail geometry permits the Building Regulation heights for guarding to be maintained without sacrificing the integrity of the geometrical relationship between flight and balustrade.

The third solution arises from traditional construction where timber strings form the supporting structure for balustrade and handrail. Such construction can be formed today in steel tubular members and glass or steel and provide a structural balustrading alongside the staircase. It can occupy a zonal plane to fulfil the guarding role and which, if space permits will provide an elegant screening to the stairs (Fig. 8.4(f) and (h)).

The provisos made above concerning dog-leg layouts and landing spaces apply equally to the layout of three- or four-turn stairs around open wells.

Clearly, stairs which are constructed between solid enclosures involve less complicated geometry since the balustrade pattern is not relevant. Handrail alignment is still a design factor and continuous geometry should be sought to avoid abrupt jumps that would be a hazard in use.

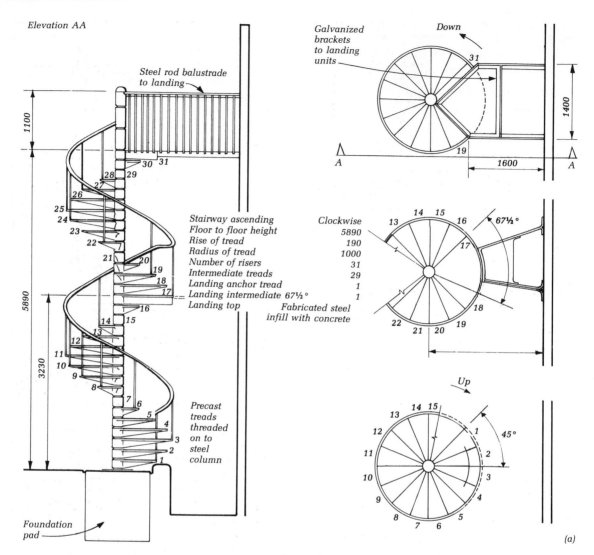

Elevation AA

1100

5890

3230

Steel rod balustrade to landing

30 31

28 29

27

26

25
24

23

22

21   20

19

18

17

16

14   15

13

12

11
10

9

8

7   6

5   4

3

2

1

Foundation pad

Stairway ascending
Floor to floor height
Rise of tread
Radius of tread
Number of risers
Intermediate treads
Landing anchor tread
Landing intermediate 67½°
Landing top

Clockwise
5890
190
1000
31
29
1
1

Fabricated steel infill with concrete

Precast treads threaded on to steel column

Galvanized brackets to landing units

Down

31

19

A   1600   A

1400

14   15
13   16
                        67½°
                                17

                                18

22   19
    21   20

Up

14 15
13   1
                    45°
12   2

11   3

10   4

9   5

8   6
    7

(a)

Figure 8.4 General points on methods of constructional design: (a) computer-aided development of spiral stairs (by kind permission of Cornish Spiral Stairs Ltd); (b) nosing alignment for dog-leg stairs with handrailing that follows comparable geometry; (c) staggered nosing alignment to save space on dog-leg stair landings; (d) traditional newel posts to overcome jumps in balustrade height; (e) deployment of handrail alignment independent of balustrade; (f) retention of kerb or string detail as basis for balustrade and handrail construction; (g) scissor-type stairs used in maisonette plans.

Economy in space often prompts designers to explore the stacking of stairs so that a single shaft accommodates two flights one above the other. Typical applications are twin escape stairs or else the interlocking of dog-leg stairs in maisonette plans. Stairs constructed within solid shafts need more space at landings if furniture has to be manoeuvred through the space, a point that is critical in 'scissors'-type plans employed for maisonette construction. (See Fig.8.4(g)).

8.4 Reinforced concrete

The structural role of reinforced concrete stairs and core walls has already been mentioned and illustrated in Fig. 8.1(d). The importance of structural geometry and the interaction between steps and balustrade or apron wall is featured in Fig. 8.4(f) and (g). Repetition of *in situ* concrete work with reused shuttering may be as economic as precasting; however, speed of installation often tilts the

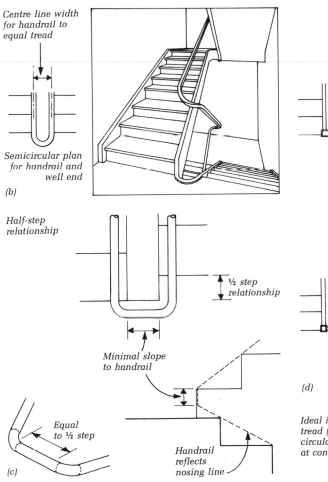

*Centre line width
for handrail to
equal tread*

*Semicircular plan
for handrail and
well end*

(b)

*Half-step
relationship*

*½ step
relationship*

*Minimal slope
to handrail*

*Equal
to ½ step*

(c)

*Handrail
reflects
nosing line*

Figure 8.4 *continued*

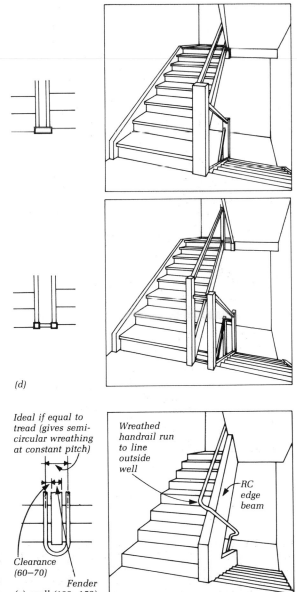

(d)

*Ideal if equal to
tread (gives semi-
circular wreathing
at constant pitch)*

*Clearance
(60–70)*

Fender

(e) wall (100–150)

*Wreathed
handrail run
to line
outside
well*

*RC
edge
beam*

choice in favour of component assembly with staircase elements which can be precast in the factory or in site workshops as Fig. 8.5(a). The break-through point in factory production are repetitive orders in excess of £150 000. The advantage of continuity in reinforcement can be exploited by casting flights with integral landings at either end. Extension bars can be arranged for casting into the surrounding concrete structure to give integrity to stairs and core construction.

The forms vary according to the structure adopted, Fig. 8.5(b)–(d) illustrating the possibilities using propped flights, spine beam construction and precast planks. The self-weight of propped slabs limits their spans to around 4.5 m unless beams can be incorporated within the balustrades or else as a central beam to the soffit with cantilevered sides. A reduction in self-weight can be made when the spine beams form the principal support to cantilever treads which are bolted in place. Such construction is more efficient in spans beyond 4.5 m and for curved staircases. Considerable accuracy is needed in assembly; the bolt and plate fastenings required between components resemble methods associated with structural steelwork. Cantilever planks in precast reinforced concrete can be made as artificial stone. These can match historic forms in turret stairs or the masonry work associated with open well helical stairs springing from a brickwork drum (see Fig. 8.1(b)). Monolithic finishes can be formed with granolithic, terrazzo or tiled surfaces to tread and riser. The general application of terrazzo permits a hard-wearing and high-quality surface to be achieved to all the exposed elements of stairs including strings. The detailed aspects of these bonded or monolithic finishes are enumerated in *MBS: Materials*. The key aspects are thicknesses and these are summarized as follows:

* *Granolithic finishing*　A thickness of 20 mm to treads,

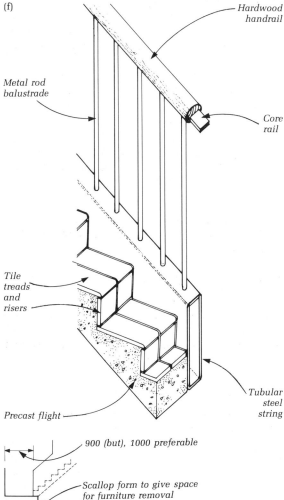

(f)

Hardwood handrail

Metal rod balustrade

Core rail

Tile treads and risers

Tubular steel string

Precast flight

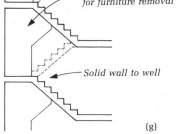

900 (but), 1000 preferable

Scallop form to give space for furniture removal

Solid wall to well

(g)

risers and strings with radius work at nosings and edges and coves at internal corners. It is usual to employ formers where overhangs occur at the nosing line (Fig. 8.5(e)).

* *Terrazzo finishing*　Requires base coat of rendering or screed of 15—20 mm thickness and then finishing with marble terrazzo coating. Minimum thicknesses of 15 mm to treads, 10 mm to risers and 6 mm for strings and skirtings depending upon size of aggregate used (Fig. 8.5(f)).
* *Tile finishing*　Tiled surfaces for treads and risers can be inserted within granolithic or terrazzo work, but the thicknesses used have to allow for bedding screeds to tread and riser elements. The advantage with terrazzo tiles is that combined units are made that combine tread and riser, with non-slip inserts for nosings (Fig.8.5(g)).

Precast spiral stairs as shown in Fig. 8.1(f) provide a unique opportunity to combine the structural and visual qualities of repetitive concrete forms. The usual assembly is made with segmental treads that are cantilevered from a central column. The design process is today automated so that the complexities incurred through the National Building Regulations of permitted ratio is part of the CAD service offered by manufacturers. The structural implications for column and cantilever construction are also synchronized for a range of materials such as steel tubular or *in situ* reinforced concrete vertical elements coupled with precast or composite treads employing concrete and steel components. Typical assembly details are given in Fig. 8.5(h) where the construction can rise four storeys as a self-sufficient structure or continuously where the landing slabs are connected to each floor deck. High-quality finishes can be applied to landing, tread and riser surfaces as already outlined for other precast flights. The repetitive forms for the cantilever units have led manufacturers to use sheet shuttering to give accurate shapes. Use of coloured or white concrete with attractive aggregates means that tooled

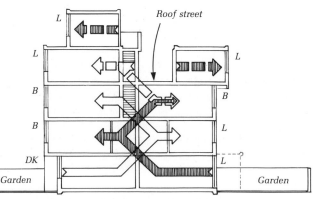

Roof street

L

L

L

B

B

B

L

DK

L

Garden

Garden

Figure 8.4　*continued*

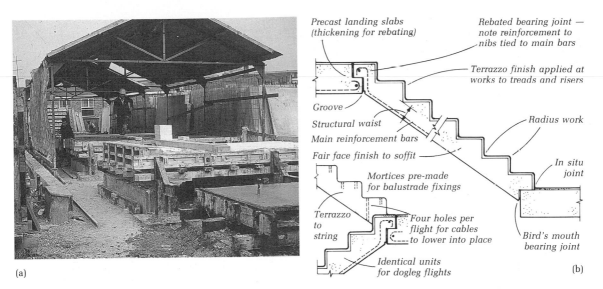

(a)

Precast landing slabs
(thickening for rebating)

Rebated bearing joint —
note reinforcement to
nibs tied to main bars

Terrazzo finish applied at
works to treads and risers

Groove

Structural waist

Main reinforcement bars

Fair face finish to soffit

Radius work

In situ
joint

Mortices pre-made
for balustrade fixings

Terrazzo
to
string

Four holes per
flight for cables
to lower into place

Bird's mouth
bearing joint

Identical units
for dogleg flights

(b)

Figure 8.5 Reinforced concrete stairs: (a) site factory for precasting stairs and landing slabs; (b) propped flights between landings; (c) spine beam construction (reprinted by kind permission of Cornish Spiral Stairs Ltd); (d) precast planks; (e) *in situ* granolithic and terrazzo finishes; (f) terrazzo tile and unit finishes; (g) combined tiles for tread and riser.

surfaces can be produced to soffits and stair edges which will match traditional masonry. This has a considerable advantage in maintenance costs over reliance upon rendering and paint, both internally and externally. Elliptical configurations are also possible.

8.5 Steel

8.5.1 Ladders and fire escapes

The traditional role for steel construction has been the fabrication of ladders and walkways for industrial buildings and the provision of external fire escapes. The basic construction involves strings made from steel flats, plate or tubular sections (Figs 8.1(g) and 8.6(a)). The treads are formed as rungs for ladder work with flats or rods, while escape stairs have step components of normal size made from steel trays or provided as fretted castings for individual treads and landings supported on angle or tubular bearers. Fretted castings are still available in cast iron. Flights can be propped between trimming beams at landings within buildings, while external escape stairs will need a framework of columns to encompass the flight (Fig. 8.6(c)). Safety provisions can include the following details and which are often part of the licence agreement made with the fire authority:

- Fire-resisting glazing where external stairs adjoin buildings.

- Rust protection to steelwork and to cast treads.
- Weather protection and perhaps defrosting elements to unprotected stairs.
- Emergency lighting.
- Security screening at exit level with doors operated by panic bolts.

The licence agreement stipulates regular inspections and requires lighting, security and steelwork to be maintained. Specialist fabricators offer a design and build service for such external stair components.

8.5.2 Folded sheet steel stairs

Industrialized production also applies to internal staircases where folded steel sheet is the principal material, often acting structurally in a composite manner with concrete screed or infilling. Balustrade fixings and edge stiffness is given by strings of steel plate, but more commonly today with tubular sections or cold-formed profiles, welded to the folded sheet (Fig. 8.6(b) and (e)). It is customary with fast-track construction in steel-framed buildings to advance the stair framing and folded steel treads so that access to the floor decking can be made without external scaffolding. The ultimate finishing to the steel stairs can depend upon encasement with other materials, for example hardwood facing, or with rendering and screed plus granolithic, terrazzo or tile finishes as for concrete construction. It is also possible to simplify the finishing by infilling the steel trays for treads and landings, the exposed strings and soffits being cleaned and spray coated as shown in Fig. 8.6(f).

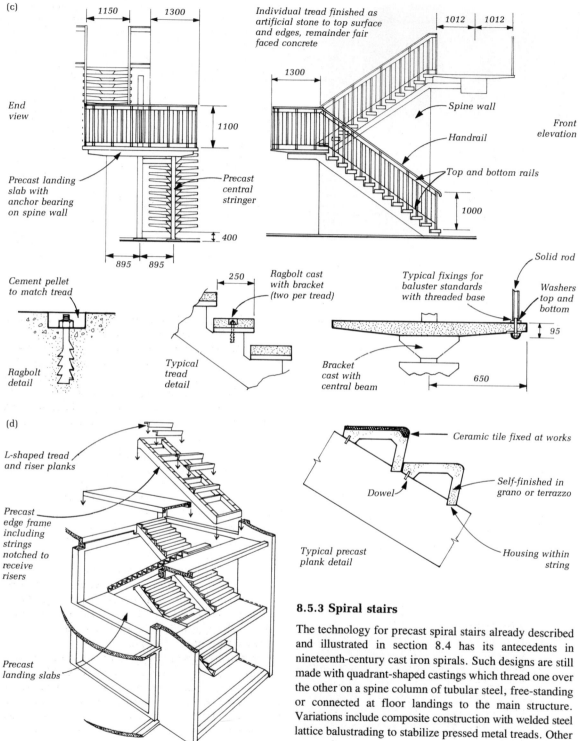

(c)

1150   1300

End
view

Individual tread finished as
artificial stone to top surface
and edges, remainder fair
faced concrete

1012   1012

1300

1100

Spine wall

Handrail

Front
elevation

Precast landing
slab with
anchor bearing
on spine wall

Precast
central
stringer

Top and bottom rails

1000

400

895   895

Cement pellet
to match tread

250

Ragbolt cast
with bracket
(two per tread)

Typical fixings for
baluster standards
with threaded base

Solid rod

Washers
top and
bottom

95

Ragbolt
detail

Typical
tread
detail

Bracket
cast with
central beam

650

(d)

L-shaped tread
and riser planks

Precast
edge frame
including
strings
notched to
receive
risers

Ceramic tile fixed at works

Dowel

Self-finished in
grano or terrazzo

Typical precast
plank detail

Housing within
string

Precast
landing slabs

Figure 8.5   *continued*

8.5.3 Spiral stairs

The technology for precast spiral stairs already described
and illustrated in section 8.4 has its antecedents in
nineteenth-century cast iron spirals. Such designs are still
made with quadrant-shaped castings which thread one over
the other on a spine column of tubular steel, free-standing
or connected at floor landings to the main structure.
Variations include composite construction with welded steel
lattice balustrading to stabilize pressed metal treads. Other
composite arrangements depend upon acrylic sheet or
toughened glass balustrades which are capped by a tubular
steel handrail. Effective connections between sheet

(e)

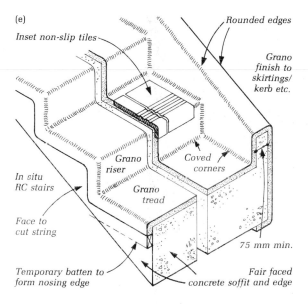

Inset non-slip tiles

Rounded edges

Grano finish to skirtings/ kerb etc.

In situ RC stairs

Grano riser

Coved corners

Grano tread

Face to cut string

Temporary batten to form nosing edge

75 mm min.

Fair faced concrete soffit and edge

(g)

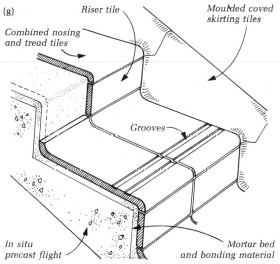

Riser tile

Moulded coved skirting tiles

Combined nosing and tread tiles

Grooves

In situ precast flight

Mortar bed and bonding material

(f)

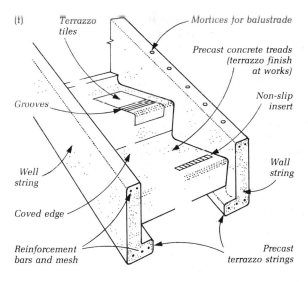

Terrazzo tiles

Mortices for balustrade

Precast concrete treads (terrazzo finish at works)

Non-slip insert

Grooves

Well string

Wall string

Coved edge

Reinforcement bars and mesh

Precast terrazzo strings

Mesh fabric reinforcement

Precast terrazzo planks with in situ terrazzo work to margins and skirting coves

Fair finish for paint or terrazzo as rest of facing material

Precasts or in situ RC stair

Figure 8.5 *continued*

balustrading and the strings extends their structural role while the flexing and tensile stresses are resolved by the upper chord of the tubular handrailing (Fig. 8.6(d)). There are many specialists in this field who offer engineering design skills to master both the dimensional criteria of the Building Regulations and structural solutions to the Code of Practice requirements. The majority of fabricators are marketing sets of parts that are interchangeable within a particular structural system.

8.5.4 Suspended and welded stairs

Independence from standardized components means a fully engineered solution with designer and engineer embarking on staircase structures beyond catalogue solutions.

The imaginative use of welded tubular steel will give the greatest freedom in design — the possibilities with cantilevers are more satisfactory in use since standard spirals depend upon bolted connections which may eventually loosen. The example illustrated in Fig. 8.6(g) reveals this flair and sinuous quality that can be developed in the Netherlands where the Building Regulations are mercifully free of restrictions.

The designs by Eva Jiricna and her engineer consultants have advanced the concept of suspended trussed stairs to an art form. The best example is the triple flight which serves the Joseph Store, Sloane Street, and illustrated in Fig. 8.6(h). Here, the engineering has been honed to the absolute minimum, and the visual qualities are enhanced by glass treads and balustrading carried by a structure which is a light assembly of stretched cables, stainless rods and connectors.

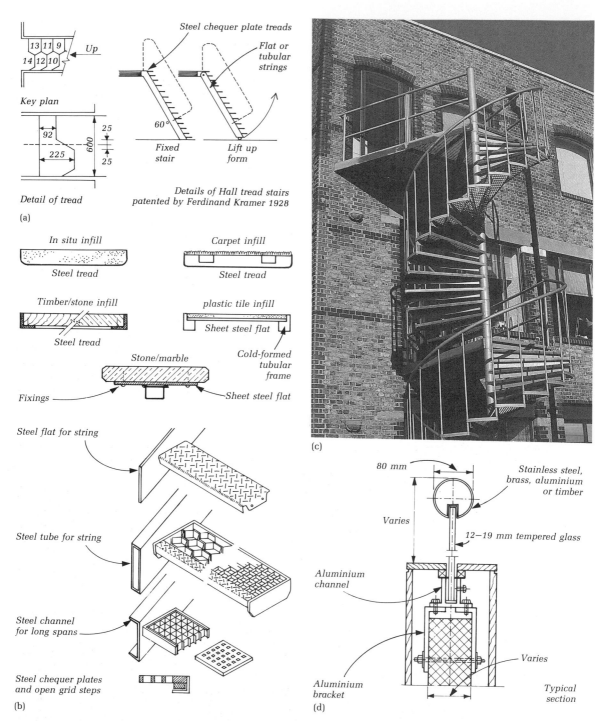

Figure 8.6 Steel staircases: (a) ladder steps; (b) typical string and tread trays; (c) external escape stairs; (d) sheet balustrading with top chord of tubular steel (reprinted by kind permission of JRL Park Royal House, 23 Park Royal Road, London NW10 7JH); (e) folded steel sheet for industrialized stairs; (f) typical installation of steel stairs in offices, Stuckley Street, London (1992) (architects: Jestico and Whiles) (photo: Jo Reid and John Peck); (g) elegantly welded tubular work for stairs and handrails, City Hall, Arnhem, the Netherlands; (h) suspended stairs, Joseph Store, Sloane Square, London (1988) architects: Eva Jiricna; engineer Matthew Wells (Whitby and Bird)).

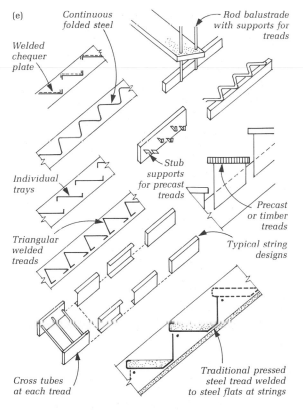

(e)

Continuous folded steel

Welded chequer plate

Rod balustrade with supports for treads

Individual trays

Stub supports for precast treads

Precast or timber treads

Triangular welded treads

Typical string designs

Cross tubes at each tread

Traditional pressed steel tread welded to steel flats at strings

8.6 Timber stairs

8.6.1 Open tread stairs (Fig. 8.7(a))

Ladders, and particularly those developed in shipbuilding, are the prototype for open tread stairs. The stability relies on effective connections between the strings in the absence of risers or a boarded soffit. The key design points are as follows:

- Straight, uncut strings to maximize longitudinal strength, usually limited to 16 risers.
- Solid plank or laminated treads plus a generous glue line to the mortices. Width of stair usually limited to 1.0 m.
- Steel cross-ties with plate bolt ends to tie the strings together, say a minimum of 3.0 m length.
- Part risers will assist support to treads and will increase the glue line. They also help with compliance concerning the maximum 100 mm gap defined in the Building Regulations.

Wider versions will require timber engineering designs with two or three beams termed 'carriage pieces' at 600 to 700 mm intervals. Strings or carriage pieces built up as laminated sections will prove more reliable than traditional 45 mm strings of solid timber in large-scale work. Blocking pieces or metal brackets will be stronger than cut strings and help reduce the volume of timber cut to waste as explained in Fig. 8.7(b).

(f)

(g)

(h)

Figure 8.6   *continued*

8.6.2 Cased staircase construction

Conventional framing depends upon 'boxed or cased' construction to provide structural integrity to the entire staircase. The flights are designed as self-contained elements propped from trimmer to trimmer or between landings, framing and newel posts. Mass production and the requirement of preferred dimensions have given rise to staircase kits offered in a range of modules with standardized detail.[3]

It is possible to assemble direct flights, dog-leg and multiple-turn layouts with standard staircases to complete the installation on site from a range of component designs for newels, balustrades, handrailing and trim for platform treads and return nosings. Figure 8.7(c) shows typical pages from a supplier of staircase kits. There are many circumstances where the historic context implies working to details drawn from the past where traditional full sizing (namely drawing out details to full size) will be needed to satisfy aesthetic standards. It is once again a matter of referring to older sources and textbooks as outlined in ref. 2.

The principles of cased construction are explained in Fig.

8.7(d) which illustrates the preparation of strings with routed work to receive risers, treads and wedges. The final glued assembly is cramped together with the underside framing reinforced by glued and screwed blocks. The nosing edge profile is dictated by the housing joint of the riser; a further complication occurs with cut strings, where the nosing profile continues from the front edge of the tread around to the exposed side of the tread. Carpet finishing in such circumstances implies painted margins and the unsightly look of a selvedged edge behind the balusters. Straight strings avoid these complications. Newel posts form a crucial element in the framing at landings by providing a bearing and connection joints for strings and handrails. They also ease the geometry of handrails by dispensing with wreathed turns (see Figs 8.4(b) (wreathed turns to handrails) and 8.3(d) (newel post framing)).

8.6.3 Spiral and curved framing

Composite construction with steel and timber exists with many proprietary spiral stairs where wooden elements are

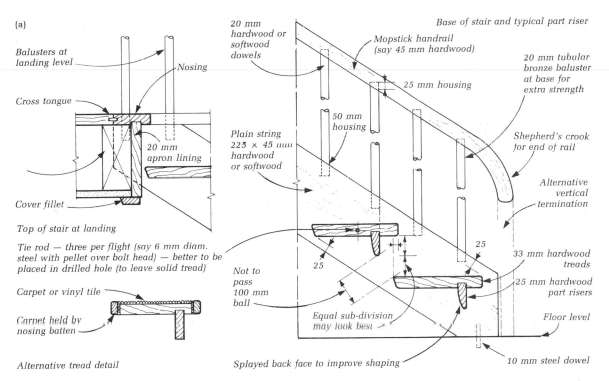

(a)

Balusters at landing level

Nosing

Cross tongue

20 mm apron lining

Cover fillet

Top of stair at landing

Tie rod — three per flight (say 6 mm diam. steel with pellet over bolt head) — better to be placed in drilled hole (to leave solid tread)

Carpet or vinyl tile

Carpet held by nosing batten

Alternative tread detail

20 mm hardwood or softwood dowels

Plain string 225 × 45 mm hardwood or softwood

Not to pass 100 mm ball

Base of stair and typical part riser

Mopstick handrail (say 45 mm hardwood)

25 mm housing

20 mm tubular bronze baluster at base for extra strength

50 mm housing

Shepherd's crook for end of rail

Alternative vertical termination

25

33 mm hardwood treads

25 mm hardwood part risers

25

Floor level

Equal sub-division may look best

Splayed back face to improve shaping

10 mm steel dowel

Figure 8.7 Timber stairs: (a) open tread stairs; (b) laminated strings with blocking pieces; (c) typical pages from a supply of kit components for cased stairs (by kind permission of Richard Burbidge Ltd); (d) principles of cased construction; (e) Georgian elliptical stairs (from James Newlands *The Carpenter's Assistant*.[2])

reserved for handrails and for facings to metal-framed treads. Timber construction for curved or spiral work involves highly specialized fabrication, the techniques dating back to the Renaissance, together with the mathematical basis for calculating the curves and setting out. Amplification of such details are found in most joinery textbooks from the nineteenth century. The balustrade, handrailing and strings often play a crucial role with mechanical joints formed by bolts secretly housed into the handrail and string. Metal balusters (square or circular bars) are also employed which match the timber profiles, but are brazed to plates at either end. Strings are usually uncut to preserve the full profile of the section and often laminated to give additional strength. Steel flitch plates (say 250 × 6 mm) can also be included within the lamination, brazed connections to occasional steel balusters providing greater rigidity to the balustrading. Graceful stairs on these principles can be found in many Georgian buildings, Fig. 8.7(e) being a typical elegant example.

Contemporary designs often revert to stairs built to flowing lines without the interruption of newel posts; see Fig. 8.7(c) and (d) for comparisons with traditional work.

Economy in space means designing with a central newel post instead of an open well. Timber is well suited to this arrangement provided the newel is of sufficient size to accommodate the glued mortices for treads and risers, say 175 mm diameter or 175 × 175 mm. It is possible to form plywood drums for this purpose, but the size will increase to 250 mm diameter or 250 mm square. The outer strings if curved can be laminated as already described. A cheaper solution is to allow for a square plan if the stair is placed within a recess and for the stair assembly to resemble winders framed by wall strings of solid timber or plywood.

8.7 Cantilever stone stairs (Fig. 8.8(a))

There has been renewed interest in building helical stairs. The use of artificial or reconstructed stone permits construction costs that are comparable to timber framing. The pattern books of Palladio refer to masonry drums that were built with cantilever stone blocks to form staircase treads and risers around an open well. Each block has a bearing within the enclosing masonry of 110—225 mm, with the risers having a rebated bearing one upon the other

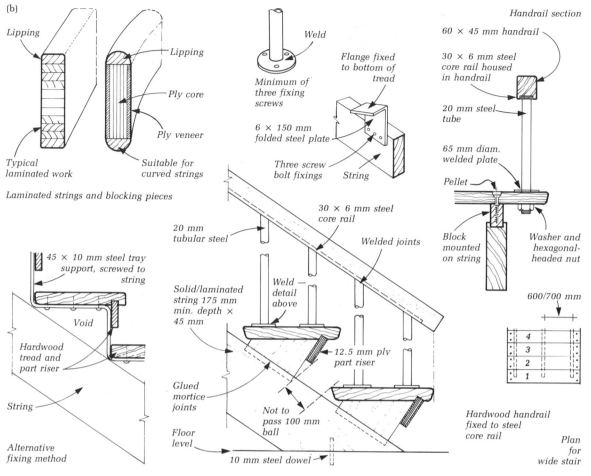

(b)

Lipping

Lipping

Ply core

Ply veneer

Typical laminated work

Suitable for curved strings

Laminated strings and blocking pieces

Weld

Minimum of three fixing screws

6 × 150 mm folded steel plate

Flange fixed to bottom of tread

Three screw bolt fixings

String

Handrail section

60 × 45 mm handrail

30 × 6 mm steel core rail housed in handrail

20 mm steel tube

65 mm diam. welded plate

Pellet

Block mounted on string

Washer and hexagonal-headed nut

600/700 mm

Hardwood handrail fixed to steel core rail

Plan for wide stair

45 × 10 mm steel tray support, screwed to string

20 mm tubular steel

30 × 6 mm steel core rail

Welded joints

Void

Solid/laminated string 175 mm min. depth × 45 mm

Weld — detail above

Hardwood tread and part riser

12.5 mm ply part riser

String

Glued mortice joints

Not to pass 100 mm ball

Alternative fixing method

Floor level

10 mm steel dowel

Figure 8.7   *continued*

in the rise of the flight. There are notable stairs by Wren on this pattern in the South-west tower to St Paul's Cathedral and within the Monument built in 1671–77 to commemorate the Great Fire of London. The construction clearly involves propping until the tread blocks and walling masonry have cured. There is also the additional bracing effect given by metal balusters which are 'leaded' into mortices of the treads with secure fixing to the metal handrail or core rail (if timber is employed as a capping). 'Leaded' work can be explained as follows. The forked ends of the iron were set into molten lead filling to holes 'morticed' into masonry. The lateral stability is ensured by the close spacing of the vertical ironwork (between 100 and 250 mm centres). The detail in Fig. 8.8(b) reveals the bearings needed in the wall masonry and between each tread and riser. In this example the comparative prices in 1990 were £22 000 for artificial stone cantilever with steel balustrade, compared with £28 000 for timber framing with curving strings, etc.

8.8 Balustrades and screens

8.8.1 General points

The preceding examples of stair designs have referred to the appropriate forms of balustrade which are likely to form the pattern continued for landings and associated balconies. These are many locations where the guarding of landings and balconies within buildings is considered separately from stair protection. The rules concerning height and security are stated within the National Building Regulations and Codes of Practice as referred to and illustrated in sections 8.2.1–8.2.3 (see Table 8.1 for loading requirements).

The constructional methods can be summarized as follows and are illustrated in Fig. 8.9(a)–(i).

8.8.2 Solid balustrades

These are often formed in reinforced concrete or precast

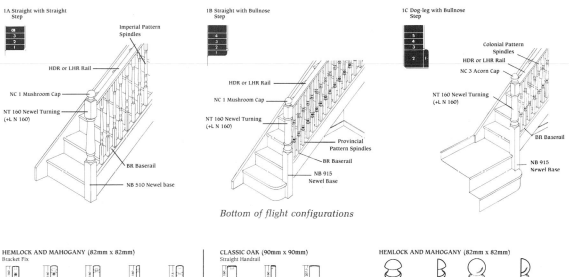

1A Straight with Straight Step

Imperial Pattern Spindles

HDR or LHR Rail

NC 1 Mushroom Cap

NT 160 Newel Turning (+L N 160)

BR Baserail

NB 510 Newel base

1B Straight with Bullnose Step

HDR or LHR Rail

NC 1 Mushroom Cap

NT 160 Newel Turning (+L N 160)

Provincial Pattern Spindles

BR Baserail

NB 915 Newel Base

1C Dog-leg with Bullnose Step

Colonial Pattern Spindles

HDR or LHR Rail

NC 3 Acorn Cap

NT 160 Newel Turning (+L N 160)

BR Baserail

NB 915 Newel Base

Bottom of flight configurations

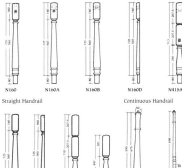

HEMLOCK AND MAHOGANY (82mm x 82mm)
Bracket Fix

N160   N160A   N160B   N160D   N415A

Straight Handrail

NT160   NT160 HALF   NT415   NT202

Continuous Handrail

NTO   NT-V

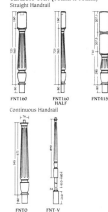

CLASSIC OAK (90mm x 90mm)
Straight Handrail

FNT160   FNT160 HALF   FNT415   FNT202

Continuous Handrail

FNTO   FNT-V

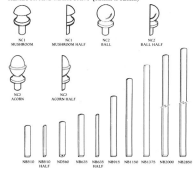

HEMLOCK AND MAHOGANY (82mm x 82mm)

NC1 MUSHROOM   NC1 MUSHROOM HALF   NC2 BALL   NC2 BALL HALF

NC3 ACORN   NC3 ACORN HALF

NB510   NB510 HALF   ND560   NB635   NB635 HALF   NB915   NB1150   NB1375   NB2000   NB2850

Spindles, newel bases and caps

Handrail fittings

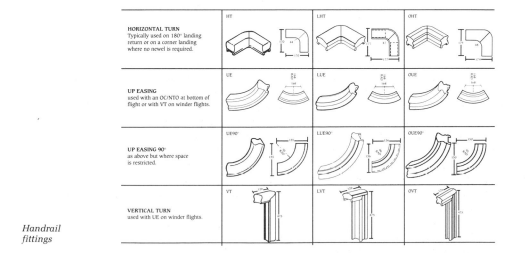

	HT	LHT	OHT
HORIZONTAL TURN Typically used on 180° landing return or on a corner landing where no newel is required.			
	UE	LUE	OUE
UP EASING used with an OC/NTO at bottom of flight or with VT on winder flights.			
	UE90°	LUE90°	OUE90°
UP EASING 90° as above but where space is restricted.			
	VT	LVT	OVT
VERTICAL TURN used with UE on winder flights.			

(c)

Figure 8.7 *continued*

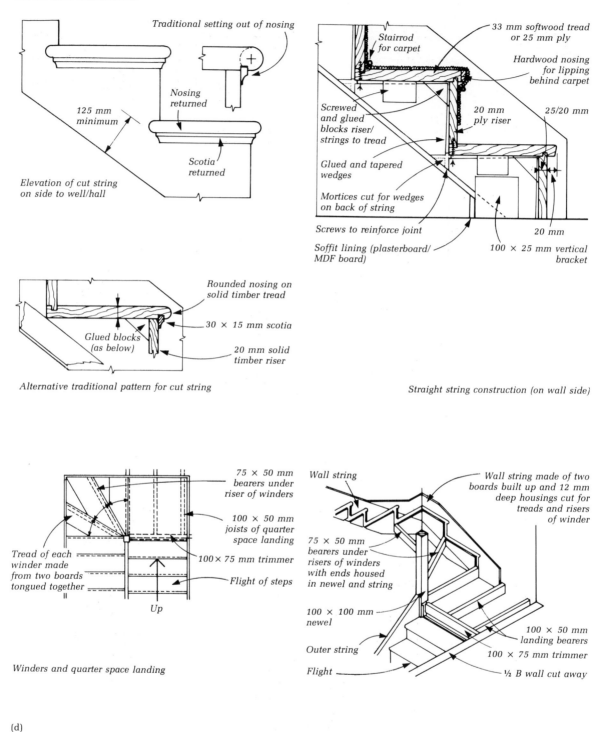

Traditional setting out of nosing

Nosing returned

125 mm minimum

Scotia returned

Elevation of cut string on side to well/hall

33 mm softwood tread or 25 mm ply

Stairrod for carpet

Hardwood nosing for lipping behind carpet

Screwed and glued blocks riser/ strings to tread

20 mm ply riser

25/20 mm

Glued and tapered wedges

Mortices cut for wedges on back of string

Screws to reinforce joint

Soffit lining (plasterboard/ MDF board)

20 mm

100 × 25 mm vertical bracket

Straight string construction (on wall side)

Rounded nosing on solid timber tread

30 × 15 mm scotia

Glued blocks (as below)

20 mm solid timber riser

Alternative traditional pattern for cut string

75 × 50 mm bearers under riser of winders

100 × 50 mm joists of quarter space landing

Tread of each winder made from two boards tongued together

100 × 75 mm trimmer

Flight of steps

Up

Winders and quarter space landing

Wall string

Wall string made of two boards built up and 12 mm deep housings cut for treads and risers of winder

75 × 50 mm bearers under risers of winders with ends housed in newel and string

100 × 100 mm newel

100 × 50 mm landing bearers

100 × 75 mm trimmer

Outer string

½ B wall cut away

Flight

(d)

Figure 8.7 *continued*

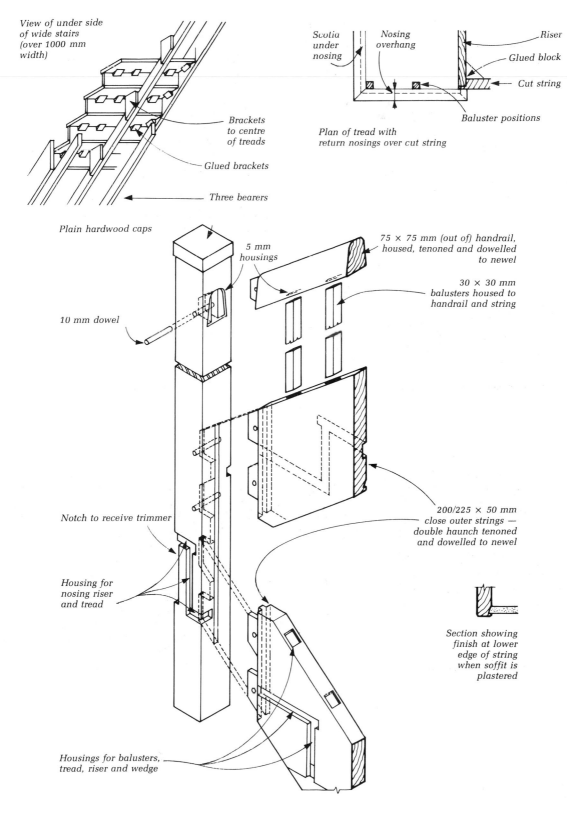

*View of under side
of wide stairs
(over 1000 mm
width)*

Brackets
to centre
of treads

Glued brackets

Three bearers

Scotia
under
nosing

Nosing
overhang

Riser

Glued block

Cut string

Baluster positions

*Plan of tread with
return nosings over cut string*

Plain hardwood caps

5 mm
housings

10 mm dowel

*75 × 75 mm (out of) handrail,
housed, tenoned and dowelled
to newel*

*30 × 30 mm
balusters housed to
handrail and string*

Notch to receive trimmer

Housing for
nosing riser
and tread

*200/225 × 50 mm
close outer strings —
double haunch tenoned
and dowelled to newel*

*Section showing
finish at lower
edge of string
when soffit is
plastered*

Housings for balusters,
tread, riser and wedge

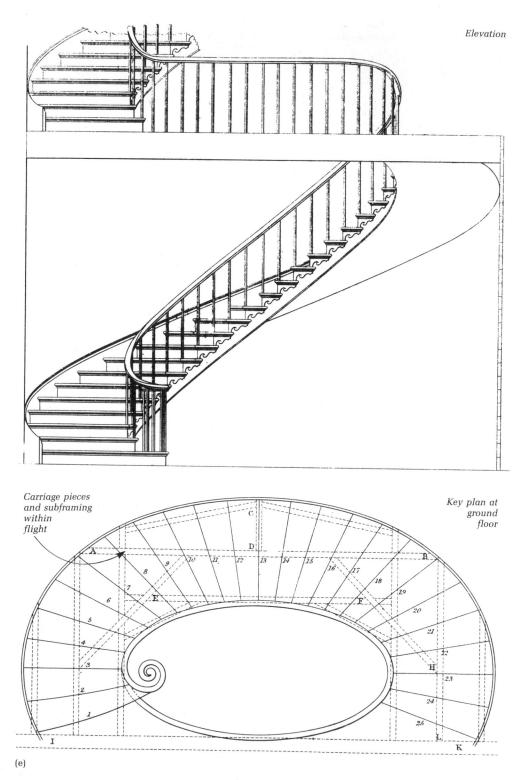

Elevation

*Carriage pieces
and subframing
within
flight*

*Key plan at
ground
floor*

(e)

Figure 8.7   *continued*

and devised as a structural component with thicknesses between 100 and 175 mm. They can also be built up with brick or blockwork but will need reinforcement to provide lateral stability. Perforated masonry permits vertical rods to be passed through the units to give continuity with landings, etc. while steel mesh can be built into bed joints to improve the overall strength of balcony construction (Fig. 8.9(a)). Handrailing will form a useful adjunct in all situations related to stairs and can be arranged as continuous lengths of tubular steel (either coated or stainless). These can be carried on side- or top-fixing brackets and placed at the appropriate height.

8.8.3 Solid kerbs to balustrades

This construction also performs a similar structural role with *in situ* or precast concrete. The kerb height of 150—300 mm improves security at balcony and stair edges by preventing the accidental fall of cleaning materials and other objects down into stair-wells or open spaces below balconies. In industrial buildings raised kerbs to balconies or landings are essential to prevent dropped tools or other materials from falling away into open spaces below floors and stairways. The kerb reduces the extent of balustrading and can provide secure fixings into the top edge for metal balusters and standards, as shown in Fig. 8.9(b). Industrialized components include terrazzo and tiled-faced elements for raised kerb construction employing precast concrete as the core material. Kerbs can also be fabricated from hollow tubular steel, as shown in Fig. 8.9(c) and (d), say 75/100 × 200/300 mm sections or with laminated timber of similar sizes, such kerbs forming a carrier for metal balusters and standards.

8.8.4 Balustrades and traditional balusters

This traditional basis for metal construction favoured wrought iron balusters 'leaded' into mortices of stone landing or stair slabs. The combined geometry of landing shape and cross-stays asserts stability as shown in Figs 8.8(a) and (b) and 8.9(e). Such constructional methods are still used for external railings and for historic replicas in helical stairs. The lack of wrought iron forging today means steel is now used exclusively for metal balustrading in rod or tube form apart from some decorative work in cast iron, aluminium or bronze.

Balusters formed with timber components using string and handrails are already described under timber stairs, the details applying to balcony and landing edges where timber construction exists.

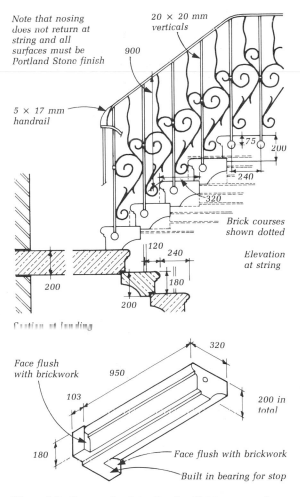

Figure 8.8 Constructional details of artificial stone treads, and walling and balustrades at Cromwell House, Highgate (architect: Russell Taylor, engineer: Sam Price, 1990).

8.8.5 Balustrades with standards and panels

This pattern is more commonly used for industrialized components since it permits standardized panels to be fitted within an overall framework of heavier structural members. The structure involves vertical standards which are engineered to take applied lateral loading from the total assembly and to safely transfer this to the floor edge or stair periphery. The standards are often tubular to save weight and improve torsional resistance. The standards have to be so arranged that the panel elements are not oversized for the spans involved, say, limited to 1500—1800 mm centres, as shown in Fig. 8.9(f). The verticals are connected lengthwise by structural handrails and by panels or by subsidiary core rails at the top and bottom of infilling

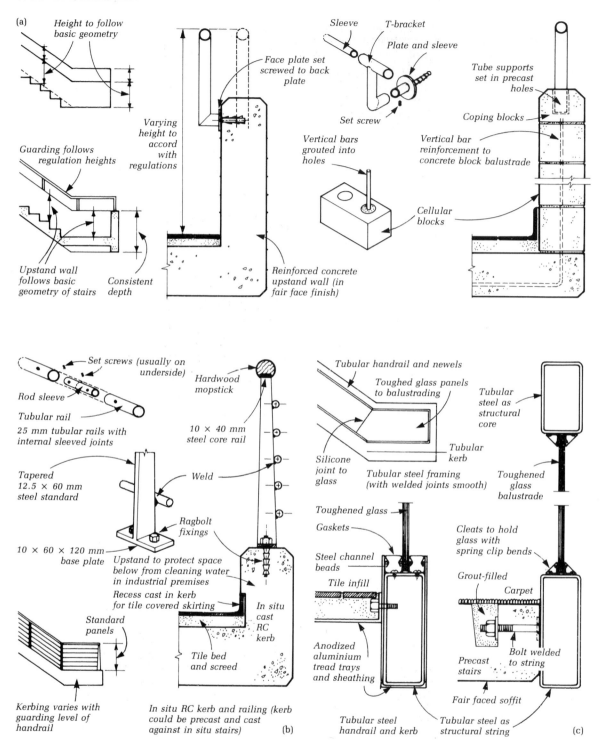

Figure 8.9 Balustrades: (a) solid balustrade; (b) solid kerb; (c) metal tubular kerb; (d) laminated timber kerb; (e) individual iron and steel baluster fixings; (f) standards with panelled balustrade; (g) alternative panel treatment with frameless glass panels (by kind permission of J. D. Beardmore); (h) fixing details for metal barriers (BS 6180 : 1982); (i) glasscrete screens.

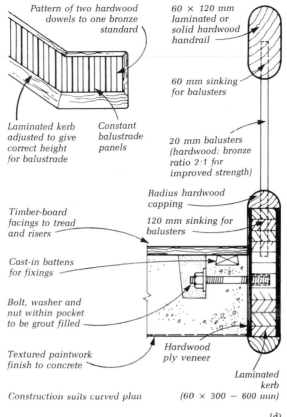

Pattern of two hardwood dowels to one bronze standard

Laminated kerb adjusted to give correct height for balustrade

Constant balustrade panels

60 × 120 mm laminated or solid hardwood handrail

60 mm sinking for balusters

20 mm balusters (hardwood: bronze ratio 2·1 for improved strength)

Radius hardwood capping

Timber-board facings to tread and risers

120 mm sinking for balusters

Cast-in battens for fixings

Bolt, washer and nut within pocket to be grout filled

Textured paintwork finish to concrete

Hardwood ply veneer

Laminated kerb (60 × 300 − 600 mm)

Construction suits curved plan

(d)

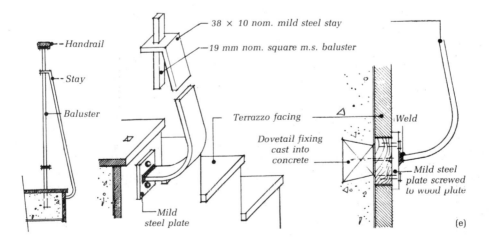

Handrail

Stay

Baluster

38 × 10 nom. mild steel stay

19 mm nom. square m.s. baluster

Terrazzo facing

Dovetail fixing cast into concrete

Weld

Mild steel plate

Mild steel plate screwed to wood plate

(e)

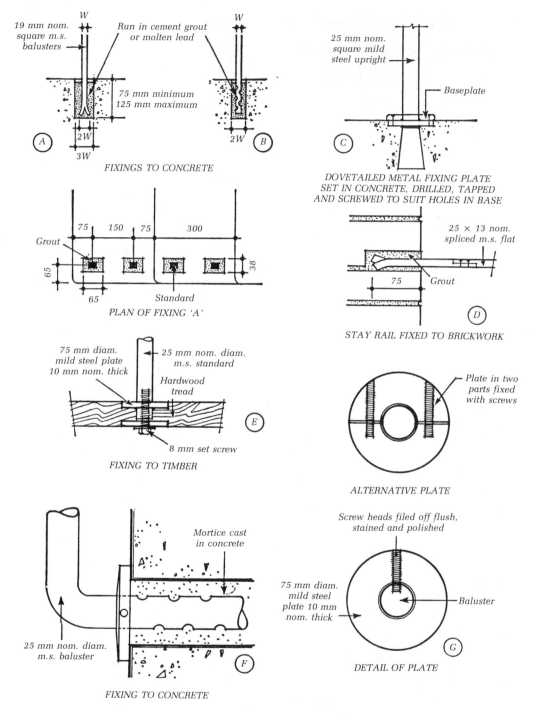

19 mm nom. square m.s. balusters

Run in cement grout or molten lead

W

75 mm minimum
125 mm maximum

2W

3W

A

W

2W

B

FIXINGS TO CONCRETE

25 mm nom. square mild steel upright

Baseplate

C

*DOVETAILED METAL FIXING PLATE
SET IN CONCRETE, DRILLED, TAPPED
AND SCREWED TO SUIT HOLES IN BASE*

75   150   75   300

Grout

65

65

Standard

38

PLAN OF FIXING 'A'

25 × 13 nom. spliced m.s. flat

75   Grout

D

STAY RAIL FIXED TO BRICKWORK

75 mm diam. mild steel plate 10 mm nom. thick

25 mm nom. diam. m.s. standard

Hardwood tread

E

8 mm set screw

FIXING TO TIMBER

Plate in two parts fixed with screws

ALTERNATIVE PLATE

Mortice cast in concrete

25 mm nom. diam. m.s. baluster

F

FIXING TO CONCRETE

Screw heads filed off flush, stained and polished

75 mm diam. mild steel plate 10 mm nom. thick

Baluster

G

DETAIL OF PLATE

(e) *cont.*

Figure 8.9 *continued*

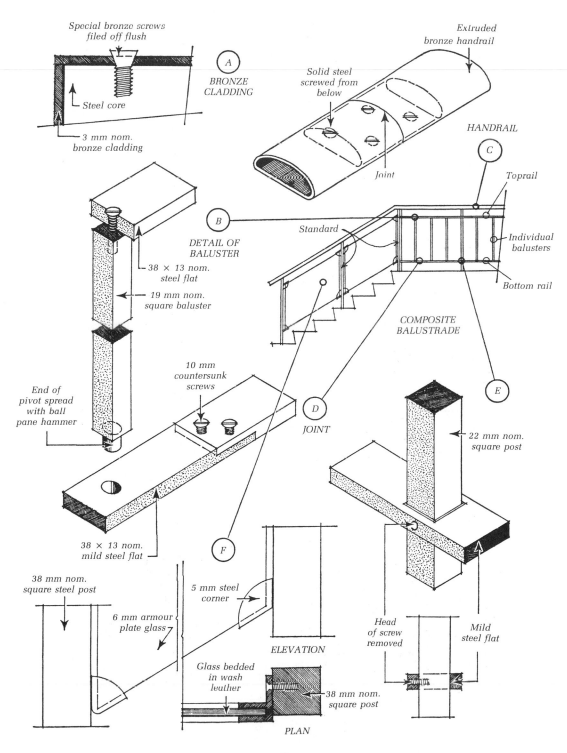

Special bronze screws
filed off flush

A

*BRONZE
CLADDING*

Steel core

3 mm nom.
bronze cladding

Extruded
bronze handrail

Solid steel
screwed from
below

HANDRAIL

C

Joint

Toprail

Individual
balusters

Bottom rail

Standard

B

*DETAIL OF
BALUSTER*

38 × 13 nom.
steel flat

19 mm nom.
square baluster

*COMPOSITE
BALUSTRADE*

E

10 mm
countersunk
screws

D

JOINT

22 mm nom.
square post

End of
pivot spread
with ball
pane hammer

38 × 13 nom.
mild steel flat

F

38 mm nom.
square steel post

5 mm steel
corner

Head
of screw
removed

Mild
steel flat

6 mm armour
plate glass

ELEVATION

Glass bedded
in wash
leather

38 mm nom.
square post

PLAN

(f)

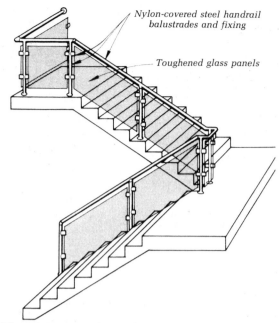

Nylon-covered steel handrail balustrades and fixing

Toughened glass panels

Figure 8.9   *continued*

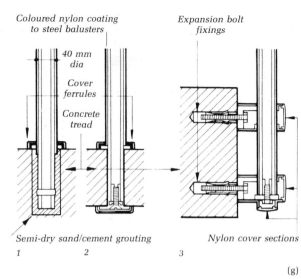

Coloured nylon coating to steel balusters

40 mm dia

Cover ferrules

Concrete tread

Expansion bolt fixings

Semi-dry sand/cement grouting

1   2   3

Nylon cover sections

(g)

materials. The total geometry of the framework plays a significant role in the engineering design to resist side-loading stresses from traffic on the balconies, landing or stairs. The main frame joints between standards and lateral members are commonly welded or made by set screwing one element to another. The infilling panels can be arranged as welded rods running between connecting rails, termed 'core rails' as shown in Fig. 8.9(f), or as welded mesh or perforated metal screens. The National Building Regulations call for non-climbable balustrades where children may be present, hence the interest in patterns that avoid horizontal bar designs, though these are often superior in visual effect (compare Fig. 8.9(b) and (d)).

8.8.6 Transparent panels

Transparent or translucent panels give a lighter appearance and provide a considerable measure of safety for childproof environments. The ability to see through a balustrade lessens the need to climb the obstruction. The material choices are noted with the following key characteristics.

Toughened glass   This is available in three- and five-ply thickness, with wiring to improve fire resistance and tinting to give improved visibility. Acid etching or engraved work will also prevent a sense of vertigo. Toughened glass can act structurally either in suspension from landing to landing or as cantilever panels firmly secured to floor/staircase edges. There are a number of proprietary forms of glazed balustrades that combine with tubular handrailing (Fig.

Table 8.1   Diagram 11 (AD K2/3, Section 3). Guarding design

Building category and location		Strength (kN/m)	Height (mm)
Single family dwellings	Stairs, landings, ramps, edges of internal floors	0.36	900 for all elements
	External balconies and edges of roof	0.74	1100
Factories and warehouses (light traffic)	Stairs, ramps	0.36	900
	Landings and edges of floor	0.36	1100
Residential, institutional, educational, office, and public buildings	All locations	0.74	900 for flights otherwise 1100
Assembly	530 mm in front of fixed seating	refer to BS 6399: part 1	800
	all other locations	refer to BS 6399: part 1	900 for flights elsewhere 1100
Retail	all locations	1.5	900 for flights otherwise 1100
All buildings except roof windows in loft extensions	at opening windows		800
	at glazing to changes of level	to provide containment	below 800

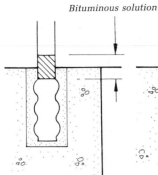
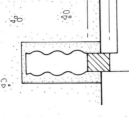

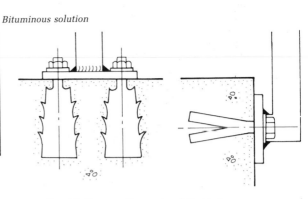

Bituminous solution   *Bituminous solution*

Standard set into-preformed or cut mortice

Standard set into preformed or cut mortice in side (pocket should be suitably reinforced)

Preset bolts normally require slotted holes, or similar tolerance in fixing plates

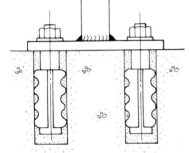
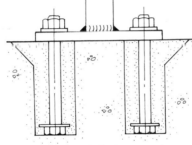
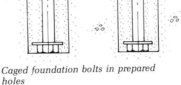
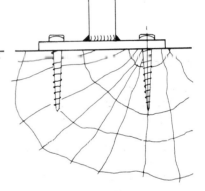

Expansion bolts in drilled holes

Caged foundation bolts in prepared holes

Coach bolts or suitable screws for fixing to timber

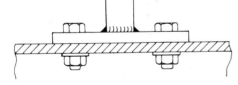
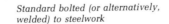

Standard bolted (or alternatively, welded) to steelwork

(h)

Figure 8.9 *continued*

8.9(g)). It should be noted that 6 mm Georgian wired plate or cast glass is no longer permitted in balustrading. Toughened glass is often used for its inherent strength rather than for its structural role.

Acrylic and polycarbonate Both materials have comparable strength to toughened glass but have the disadvantage of softening under fire conditions while acrylic has a low flash point. Polycarbonate has a superior performance if wire reinforced. Both plastics scratch readily.

Glasscrete 'Glasscrete' is a term used to describe structural assemblies of glass blocks with reinforced mortar joints or with precast glazing bars. It can be wire reinforced in both directions and constructed to be 30 minute fire resisting in panels up to 10 m². The panels will require edge restraint with steel or hardwood framing that has an integral structure with the landing or stair periphery, as shown in Fig. 8.9(i). Typical glass block sizes are 150 × 150 up to 250 × 250 mm and have thicknesses ranging from 55 to 100 mm. The material can be used in panel form within a conventional balustrade system.

(i)

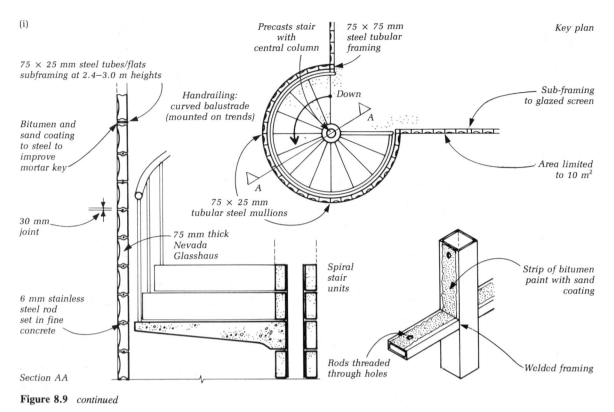

75 × 25 mm steel tubes/flats
subframing at 2.4–3.0 m heights

*Precasts stair
with
central column*

*75 × 75 mm
steel tubular
framing*

Key plan

*Handrailing:
curved balustrade
(mounted on trends)*

Down

*Sub-framing
to glazed screen*

*Bitumen and
sand coating
to steel to
improve
mortar key*

A

*Area limited
to 10 m²*

*30 mm
joint*

*75 × 25 mm
tubular steel mullions*

A

*75 mm thick
Nevada
Glasshaus*

*Spiral
stair
units*

*Strip of bitumen
paint with sand
coating*

*6 mm stainless
steel rod
set in fine
concrete*

*Rods threaded
through holes*

Welded framing

Section AA

Figure 8.9   *continued*

8.9 Case study: staircases, Japanese University College, University of Kent at Canterbury

Architects: Howell, Killick, Partridge and Amis.

The concluding illustrations (Fig. 8.10) are taken from a set of working details prepared by the architect for internal staircases serving a hall of residence at the University of Kent. The construction involves dogleg flights of open tread stairs within an independent steel angle framework with full-height newels. The framing is made as light as possible to minimize the amount of steel and timber employed and gives an open effect to the main entries. The repetitive geometry of part risers gives an attractive geometry whether viewed from above or below; the three dimensional relationship between sloping, horizontal and vertical planes are well resolved within a compact plan.

Notes

1   For *fire regulations* the most authoritative guidance can be obtained from the Institute of Building Control in their current publication: Simon Polly *Building Regulation Consulting Service: Guidance Document to the 1991 Building Regulations.* The situation is made more difficult by the impasse with the draft Fire Precautions (Place of Work) Regulations due for implementation on 1 January 1993, but withdrawn at the last moment by the Home Office. The official estimates for the costs of compliant construction were £1.7 billion. A new pared-down version is being considered that may reach the statute-book in 1994 or 1995.

2   Old textbooks and reference sources include: James Newlands *The Carpenter's Assistant*, first published by Blackie & Son Ltd in the 1850s, facsimile edition printed in 1990 by Studio Editions Ltd; W. H. Godfrey, *The English Staircase*, 1911; Ellis, *Modern Practical Stair Building and Handrailing*, (historic textbook, publisher unknown); A. Swan *The British Architect or the Builder's Treasury of Staircases*, 1775; George A. Mitchell, *Building Construction and Drawing*, Parts 1 and 2 based upon work originally compiled by Charles F. Mitchell. The 14th edition published by Batsford in 1942 still retains a considerable volume of nineteenth century detail.

3   In the UK a few manufacturers are producing standardized detail for all joinery outlets. A typical specialist is Messrs Richard Burbidge Ltd whose catalogue of components is generally available.

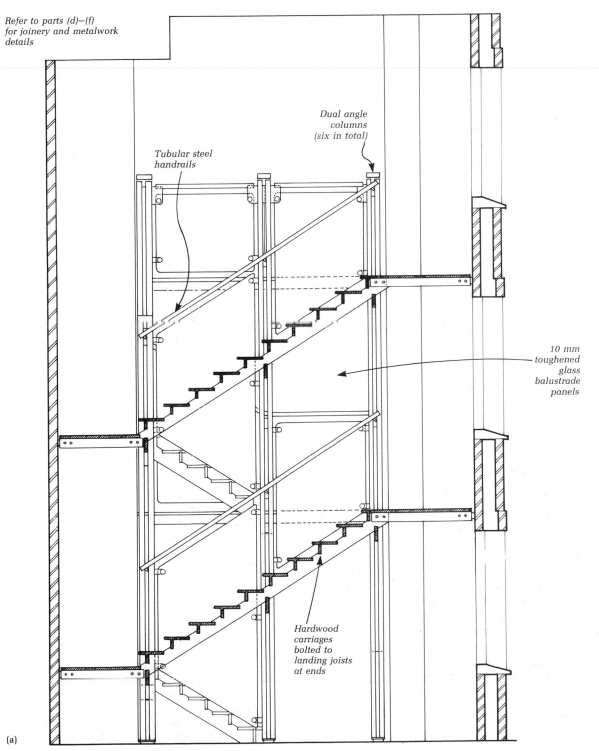

*Refer to parts (d)–(f)
for joinery and metalwork
details*

*Dual angle
columns
(six in total)*

*Tubular steel
handrails*

*10 mm
toughened
glass
balustrade
panels*

*Hardwood
carriages
bolted to
landing joists
at ends*

(a)

Figure 8.10 Staircases, Japanese University College, University of Kent at Canterbury. Working details: (a) sectional view; (b) end elevation; (c) plan of stairs and framing; (d) plan of steel angle newel; (e) detail of handrail and bracket; (f) typical tread-to-riser relationship and landing edge. (Redrawn by the kind permission of the architects.)

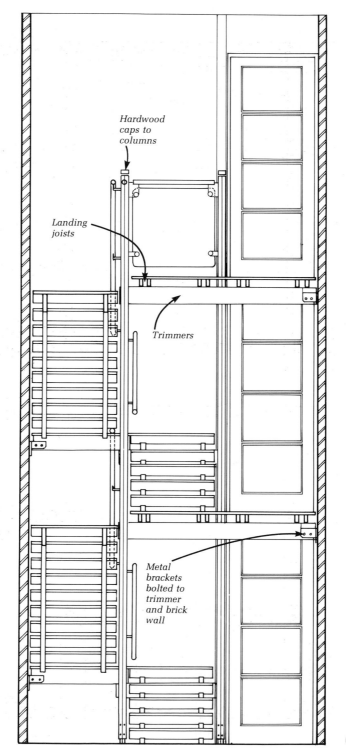

Figure 8.10 *continued*

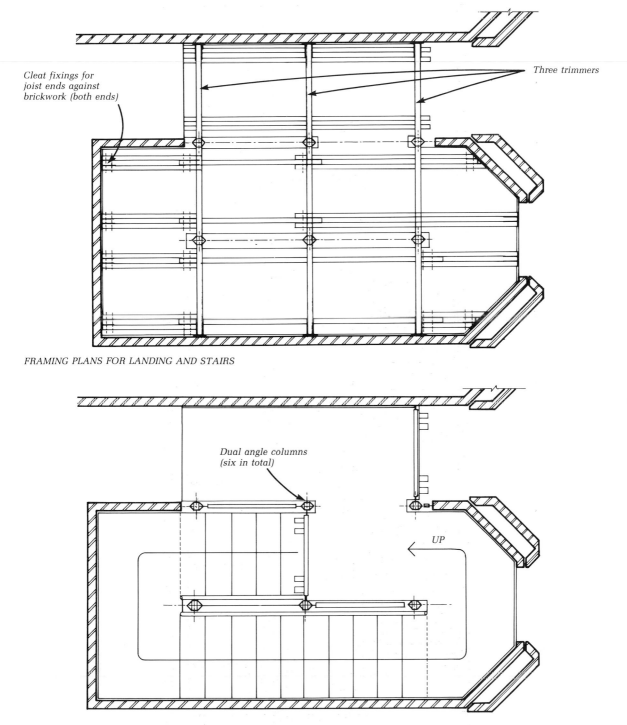

Cleat fixings for
joist ends against
brickwork (both ends)

Three trimmers

FRAMING PLANS FOR LANDING AND STAIRS

Dual angle columns
(six in total)

UP

UPPER LANDING

(c)

Figure 8.10 *continued*

(d)   *Plan details of dual angle column,*
       balustrade panel and handrail bracket

(e)   *Elevation of handrail bracket*

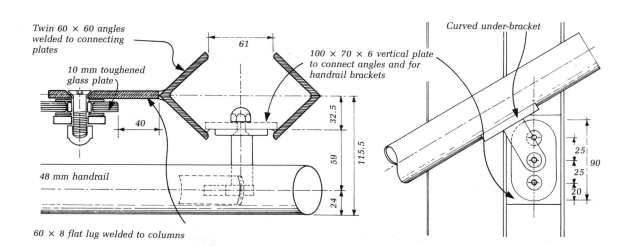

Twin 60 × 60 angles
welded to connecting
plates

10 mm toughened
glass plate

100 × 70 × 6 vertical plate
to connect angles and for
handrail brackets

61

32.5

115.5

59

24

40

48 mm handrail

60 × 8 flat lug welded to columns

Curved under-bracket

25

25

90

20

(f)   *Typical detail section*

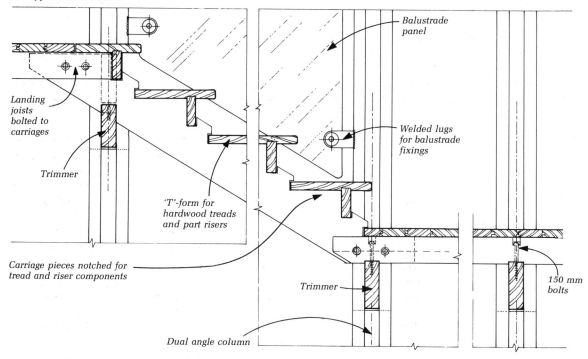

Balustrade
panel

Landing
joists
bolted to
carriages

Trimmer

'T'-form for
hardwood treads
and part risers

Welded lugs
for balustrade
fixings

Carriage pieces notched for
tread and riser components

Trimmer

150 mm
bolts

Dual angle column

Figure 8.10 *continued*

SI units

Quantities in this volume are given in SI units which have been adopted by the construction industry in the United Kingdom. Twenty-five other countries (not including the USA or Canada) have also adopted the SI system although several of them retain the old metric system as an alternative. There are six SI basic units. Other units derived from these basic units are rationally related to them and to each other. The international adoption of the SI will remove the necessity for conversions between national systems. The introduction of metric units gives an opportunity for the adoption of modular sizes.

Multiples and sub-multiples of SI units likely to be used in the construction industry are as follows:

Multiplication factor	Prefix	Symbol
1 000 000	10^6 mega	M
1 000	10^3 kilo	k
100	10^2 hecto	h
10	10^1 deca	da
0.1	10^{-1} deci	d
0.01	10^{-2} centi	c
0.001	10^{-3} milli	m
0.000 001	10^{-6} micro	μ

Note: further information concerning metrication is contained in BS PD 6031 *A Guide for the use of the Metric System in the Construction Industry*, and BS 5555: 1976 *SI units and recommendations for the use of their multiples and of certain other units.*

Quantity	Unit	Symbol	Imperial unit × Conversion factor = SI value		
LENGTH	kilometre	km	1 mile	=	1.609 km
	metre	m	1 yard	=	0.914 m
			1 foot	=	0.305 m
	millimetre	mm	1 inch	=	25.4 mm
AREA	square kilometre	km^2	1 mile^2	=	2.590 km^2
	hectare	ha	1 acre	=	0.405 ha
	square metre	m^2	1 yard^2	=	0.836 m^2
			1 foot^2	=	0.093 m^2
	square millimetre	mm^2	1 inch^2	=	645.16 mm^2
VOLUME	cubic metre	m^3	1 yard^3	=	0.765 m^3
			1 foot^3	=	0.028 m^3
	cubic millimetre	mm^3	1 inch^3	=	1 638.7 mm^3
CAPACITY	litre	1	1 UK gallon	=	4.546 litres

Continued overleaf

Quantity	Unit	Symbol	Imperial unit × Conversion factor = SI value		
MASS	kilogramme	kg	1 lb	=	0.454 kg
	gramme	g	1 oz	=	28.350 g
			1 lb/ft(run)	=	1.488 kg/m
			1 lb/ft^2	=	4.882 kg/m^2
DENSITY	kilogramme per cubic metre	kg/m^3	1 lb/ft^3	=	16.019 kg/m^3
FORCE	newton	N	1 lbf	=	4.448 N
			1 tonf	=	9 964.02 N
				=	9.964 kN
PRESSURE, STRESS	newton per square metre	N/m^2	1 lbf/in^2	=	6 894.8 N/m^2
	meganewton per	MN/m^2† or			
	square metre	N/mm^2	1 tonf/ft^2	=	107.3 kN/m^2
			1 tonf/in^2	=	15.444 MN/m^2
			1 lb/ft run	=	14.593 N/m
			1 lb/ft^2	=	47.880 N/m^2
			1 ton/ft run	=	32 682 kN/m
	*bar (0.1 MN/m^2)	bar			
	*hectobar (10 MN/m^2)	h bar			
	*millibar (100 MN/m^2)	m bar			
VELOCITY	metre per second	m/s	1 mile/h	=	0.447 m/s
FREQUENCY	cycle per second	Hz	1 cycle/sec	=	1Hz
ENERGY, HEAT	joule	J	1 Btu	=	1 055.06 J
POWER, HEAT FLOW RATE	watts	W	1 Btu/h	=	0.293 W
	newtons metres per second	Nm/s	1 hp	=	746 W
	joules per second	J/s	1 ft/1bf	=	1.356 J
THERMAL CONDUCTIVITY (k)	watts per metre degree Celsius	W/m deg C	1 Btu in/ft^{2}h deg F	=	0.144 W/m deg C
THERMAL TRANSMITTANCE (U)	watts per square metre degree Celsius	W/m^2 deg C	1 Btu/ft^{2}h deg F	=	5.678 W/m^2 deg C
TEMPERATURE	degree Celsius (difference)	° C	1° F	=	$\frac{5}{9}$ ° C
	degree Celsius (level)	° C	° F	=	$\frac{2}{5}$ ° C + 32

* Alternative units, allied to the SI, which will be encountered in certain industries
† BSI preferred symbol

CI/SfB

The following information from the *Construction Indexing Manual 1976* is reproduced by courtesy of RIBA Publications Ltd.

Used sensibly and in appropriate detail, as explained in the manual, the CI/SfB system of classification facilitates filing and retrieval of information. It is useful in technical libraries, in specifications and on working drawings. *The National Building Specification* is based on the system, and BRE Digest 172 describes its use for working drawings.

The CI/SfB system comprises tables 0 to 4, tables 1 and 2/3 being the codes in most common use. For libraries, classifications are built up from:

Table 0	Table 1	Table 2/3	Table 4
-a number code	-a number code in brackets	-upper and lower case letter codes	-upper case letter code in brackets
eg 6	eg (6)	eg Fg	eg (F)

An example for clay brickwork in walls is: (21) Fg2, which for trade literature, would be shown in a reference box as:

```
┌─────────────────────────────────────┐
│ CI/SfB 1976 reference by SfB Agency  │
│        (21)    Fg2                   │
├─────────────────────────────────────┤
│                                      │
└─────────────────────────────────────┘
```

The lower space is intended for UDC (Universal decimal classification) codes — see BS 1000A 1961. Advice in classification can be obtained from the SfB Agency UK Ltd at 39 Moreland Street, London EC1V 8BB.

In the following summaries of the five tables, chapter references are made to the six related volumes and chapters of *Mitchell's Building Series* in which aspects of the classifications are dealt with. The following abbreviations are used:

Environment and Services	ES
Materials	M
Structure and Fabric, Part 1	SF (1)
Structure and Fabric, Part 2	SF (2)
External Components	EC
Internal Components	IC
Finishes	F

Table 0 Physical Environment
(main headings only)

Scope: End results of the construction process

0	Planning areas
1	Utilities, civil engineering facilities
2	Industrial facilities
3	Administrative, commercial, protective service facilities
4	Health, welfare facilities
5	Recreational facilities
6	Religious facilities
7	Educational, scientific, information facilities
8	Residential facilities
9	Common facilities, other facilities

Table 1 Elements

Scope: Parts with particular functions which combine to make the facilities in table 0

(0-)	**Sites, projects** **Building plus external works** **Building systems** *IC 2*
(1-)	**Ground, substructure**
(11)	Ground *SF(1)* 4, 8, 11; *SF(2)* 2, 3, 11
(12)	Vacant
(13)	Floor beds *SF(1)* 4, 8; *SF(2)* 3
(14), (15)	Vacant
(16)	Retaining walls, foundations *SF(1)* 4; *SF (2)* 3, 4
(17)	Pile foundations *SF(1)* 4; *SF(2)* 3, 11
(18)	Other substructure elements
(19)	Parts, accessories, cost summary, etc
(2-)	**Structure, primary elements, carcass**
(21)	Walls, external walls *SF(1)* 1, 5; *SF (2)* 4, 5, 10
(22)	Internal walls, partitions *SF(1)* 5; *SF(2)* 4, 10; *IC 7*
(23)	Floors, galleries *SF(1)* 8; *SF(2)* 6, 10
(24)	Stairs, ramps *SF(1)* 10; *SF(2)* 8, 10
(25), (26)	Vacant
(27)	Roofs *SF(1)* 1, 7; *SF(2)* 9, 10
(28)	Building frames, other primary elements *SF(1)* 1, 6; *SF(2)* 5, 10 Chimneys *SF(1)* 9
(29)	Parts, accessories, cost summary, etc
(3-)	**Secondary elements, completion of structure**
(31)	Secondary elements to external walls, including windows, doors *SF(1)* 5; *SF(2)* 10; *EC* 2, 4, 5, 7
(32)	Secondary elements to internal walls, partitions including borrowed lights and doors *SF(2)* 10; *EC* 2, 3
(33)	Secondary elements to floors *SF(2)* 10
(34)	Secondary elements to stairs including balustrades *EC* 5
(35)	Suspended ceilings *IC 8*
(36)	Vacant
(37)	Secondary elements to roofs, including roof lights, dormers *SF(1)* 7; *SF(2)* 10; *EC* 4
(38)	Other secondary elements
(39)	Parts, accessories, cost summary, etc.
(4-)	**Finishes to structure**
(41)	Wall finishes, external *SF(2)* 4, 10; *F* 3, 4, 5

(42)	Wall finishes, internal *F* 2, 4, 5
(43)	Floor finishes *F* 1
(44)	Stair finishes *F* 1
(45)	Ceiling finishes *F* 2
(46)	Vacant
(47)	Roof finishes *SF(2)* 10; *F* 7
(48)	Other finishes
(49)	Parts, accessories, cost summary, etc
(5-)	**Services** (mainly piped and ducted)
(51)	Vacant
(52)	Waste disposal, drainage *ES* 13/*ES* 11, 12
(53)	Liquids supply *ES* 9, 10; *SF(1)* 9; *SF(2)* 6, 10
(54)	Gases supply
(55)	Space cooling
(56)	Space heating *ES* 7; *SF(1)* 9; *SF(2)* 6
(57)	Air conditioning, ventilation *ES* 7; *SF(2)* 10
(58)	Other piped, ducted services
(59)	Parts, accessories, cost summary, etc Chimney, shafts, flues, ducts independent *SF(2)* 7
(6-)	**Services** (mainly electrical)
(61)	Electrical supply
(62)	Power *ES* 14
(63)	Lighting *ES* 8
(64)	Communications *ES* 14
(65)	Vacant
(66)	Transport *ES* 15
(67)	Vacant
(68)	Security, control, other services
(69)	Parts, accessories, cost summary, etc
(7-)	**Fittings** with subdivisions (71) to (79)
(74)	Sanitary, hygiene fittings *ES* 10
(8-)	**Loose furniture, equipment** with subdivisions (81) to (89) Used where the distinction between loose and fixed fittings, furniture and equipment is important.
(9-)	**External elements, other elements**
(90)	External works, with subdivisions (90.1) to (90.8)
(98)	Other elements
(99)	Parts, accories etc. common to two or more main element divisions (1-) to (7-) Cost summary

Note: The SfB Agency UK do not use table 1 in classifying manufacturers' literature

Table 2 Constructions, Forms

Scope: Parts of particular forms which combine to make the elements in table 1. Each is characterised by the main product of which it is made.

A   Constructions, forms — used in specification applications for Preliminaries and General conditions
B   Vacant — used in specification applications for demolition, underpinning and shoring work
C   Excavation and loose fill work
D   Vacant
E   Cast *in situ* work *M* 8; *SF(1)* 4, 7, 8; *SF(2)* 3, 4, 5, 6, 8, 9

Blocks

F   Blockwork, brickwork
     Blocks, bricks *M* 6, 12; *SF(1)* 5, 9; *SF(2)* 4, 6, 7
G   Large block, panel work
     Large blocks, panels *SF(2)* 4

Sections

H   Section work
     Sections *M* 9, *SF(1)* 5, 6, 7, 8; *SF(2)* 5, 6
I   Pipework
     Pipes *SF(1)* 9; *SF(2)* 7
J   Wire work, mesh work
     Wires, meshes
K   Quilt work
     Quilts
L   Flexible sheet work (proofing)
     Flexible sheet work (proofing) *M* 9, 11
M   Malleable sheet work
     Malleable sheets *M* 9
N   Rigid sheet overlap work
     Rigid sheets for overlappings *SF(2)* 4; *F* 7
P   Thick coating work *M* 10, 11; *SF(2)* 4; *F* 1, 2, 3, 7
Q   Vacant
R   Rigid sheet work
     Rigid sheets *M* 3, 12, 13; *SF(2)* 4; *EC* 7
S   Rigid tile work
     Rigid tiles *M* 4, 12, 13; *F* 1, 4
T   Flexible sheet and tile work
     Flexible sheets eg carpets, veneers, papers, tiles cut from them *M* 3, 9; *F* 1, 6
U   Vacant
V   Film coating and impregnation work *F* 6; *M* 2
W   Planting work
     Plants

X   Work with components
     Components *SF(1)* 5, 6, 7, 8, 10; *SF(2)* 4; *IC* 5, 6; *EC* 2, 3, 4, 5, 6, 7
Y   Formless work
     Products
Z   Joints, where described separately

Table 3 Materials

Scope: Materials which combine to form the products in table 2

a   **Materials**
b,c,d   Vacant

Formed materials e to o
e   **Natural stone** *M* 4; *SF(1)* 5, 10; *SF(2)* 4
e1   Granite, basalt, other igneous
e2   Marble
e3   Limestone (other than marble)
e4   Sandstone, gritstone
e5   Slate
e9   Other natural stone

f   **Precast with binder** *M* 8; *SF(1)* 5, 7, 8, 9, 10; *SF(2)* 4 to 9; *F* 1
f1   Sandlime concrete (precast)
     Glass fibre reinforced calcium silicate (gres)
f2   All-in aggregate concrete (precast) *M* 8
     Heavy concrete (precast) *M* 8
     Glass fibre reinforced cement (gre) *M* 10
f3   Terrazzo (precast) *F* 1
     Granolithic (precast)
     Cast/artificial/reconstructed stone
f4   Lightweight cellular concrete (precast) *M* 8
f5   Lightweight aggregate concrete (precast) *M*8
f6   Asbestos based materials (preformed) *M* 10
f7   Gypsum (preformed) *EC* 2
     Glass fibre reinforced gypsum *M* 10
f8   Magnesia materials (preformed)
f9   Other materials precast with binder

g   **Clay (Dried, Fired)** *M* 5; *SF(1)* 5, 9, 10; *SF(2)* 4, 6, 7
g1   Dried clay eg pisé de terre
g2   Fired clay, vitrified clay, ceramics
     Unglazed fired clay eg terra cotta
g3   Glazed fired clay eg vitreous china
g6   Refractory materials eg fireclay
g9   Other dried or fired clays

h   **Metal** *M 9*; *SF(1)* 6, 7, *SF(2)* 4, 5, 7
h1   Cast iron
         Wrought iron, malleable iron
h2   Steel, mild steel
h3   Steel alloys eg stainless steel
h4   Aluminium, aluminium alloys
h5   Copper
h6   Copper alloys
h7   Zinc
h8   Lead, white metal
h9   Chromium, nickel, gold, other metals, metal alloys

i   **Wood** including wood laminates *M 2, 3*; *SF(1)* 5 to 8, 10; *SF(2)* 4, 9; *EC 2*
i1   timber (unwrot)
i2   Softwood (in general, and wrot)
i3   Hardwood (in general, and wrot)
i4   Wood laminates eg plywood
i5   Wood veneers
i9   Other wood materials, except wood fibre boards, chipboards and wood-wool cement

j   **Vegetable and animal materials** — including fibres and particles and materials made from these
j1   Wood fibres eg building board *M 3*
j2   Paper *M 9, 13*
j3   Vegetable fibres other than wood eg flaxboard *M 3*
j5   Bark, cork
j6   Animal fibres eg hair
j7   Wood particles eg chipboard *M 3*
j8   Wood-wool cement *M 3*
j9   Other vegetable and animal materials

k, l   Vacant

m   **Inorganic fibres**
m1   Mineral wool fibres *M 10*; *SF(2)* 4, 7
         Glass wool fibres *M 10, 12*
         Ceramic wool fibres
m2   Asbestos wool fibres *M 10*
m9   Other inorganic fibrous materials eg carbon fibres *M 10*

n   **Rubber, plastics, etc**
n1   Asphalt (preformed) *M 11*; *F 1*
n2   Impregnated fibre and felt eg bituminous felt *M 11*; *F 7*
n4   Linoleum *F 1*

         Synthetic resins n5, n6
n5   Rubbers (elastomers) *M 13*

n6   Plastics, including synthetic fibres *M 13*
         Thermoplastics
         Thermosets
n7   Cellular plastics
n8   Reinforced plastics eg grp, plastics laminates

o   **Glass** *M 12*; *SF(2)* 4; *EC 3*
o1   Clear, transparent, plain glass
o2   Translucent glass
o3   Opaque, opal glass
o4   Wired glass
o5   Multiple glazing
o6   Heat absorbing/rejecting glass
         X-ray absorbing/rejecting glass
         Solar control glass
o7   Mirrored glass, 'one-way' glass
         Anti-glare glass
o8   Safety glass, toughened glass
         Laminated glass, security glass, alarm glass
09   Other glass, including, cellular glass

Formless materials p to s
p   **Aggregates, loose fills** *M 8*
p1   Natural fills, aggregates
p2   Artificial aggregates in general
p3   Artificial granular aggregates (light) eg foamed blast furnace slag
p4   Ash eg pulverised fuel ash
p5   Shavings
p6   Powder
p7   Fibres
p9   Other aggregates, loose fills

q   **Lime and cement binders, mortars, concretes**
q1   Lime (calcined limestones), hydrated lime, lime putty, *M 7*
         Lime-sand mix (coarse stuff)
q2   Cement, hydraulic cement eg Portland cement *M 7*
q3   Lime-cement binders *M 15*
q4   Lime-cement-aggregate mixes
         Mortars (ie with fine aggregates) *M 15*; *SF(2)* 4
         Concretes (ie with fine and/or coarse aggregates) *M 8*
q5   Terrazzo mixes and in general *F 1*
         Granolithic mixes and in general *F 1*
q6   Lightweight, cellular, concrete mixes and in general *M 8*
q9   Other lime-cement-aggregate mixes eg asbestos cement mixes *M 10*

r   Clay, gypsum, magnesia and plastics binders, mortars

r1   Clay mortar mixes, refractory mortar

r2   Gypsum, gypsum plaster mixes

r3   Magnesia, magnesia mixes *F* 1

r4   Plastics binders
Plastics mortar mixes

r9   Other binders and mortar mixes

s   Bituminous materials M 11; *SF(2)* 4

s1   Bitumen including natural and petroleum bitumens, tar, pitch, asphalt, lake asphalt

s4   Mastic asphalt (fine or no aggregate), pitch mastic

s5   Clay-bitumen mixes, stone bitumen mixes (coarse aggregate)
Rolled asphalt, macadams

s9   Other bituminous materials

Functional materials t to w

t   Fixing and jointing materials

t1   Welding materials *M* 9; *SF(2)* 5

t2   Soldering materials *M* 9

t3   Adhesives, bonding materials *M* 14

t4   Joint fillers eg mastics, gaskets *M* 16 *SF(1)* 2

t6   Fasteners, 'builders ironmongery'
Anchoring devices eg plugs
Attachment devices eg connectors *SF(1)* 6, 7
Fixing devices eg bolts, *SF(1)* 5

t7   'Architectural ironmongery' *IC* 7

t9   Other fixing and jointing agents

u   Protective and Process/property modifying materials

u1   Anti-corrosive materials, treatments *F* 6
Metallic coatings applied by eg electroplating *M* 9
Non-metallic coatings applied by eg chemical conversion

u2   Modifying agents, admixtures eg curing agents *M* 8
Workability aids *M* 8

u3   Materials resisting specials forms of attack such as fungus, insects, condensation *M* 2

u4   Flame retardants if described separately *M* 1

u5   Polishes, seals, surface hardeners *F* 1; *M* 8

u6   Water repellants, if described separately

u9   Other protective and process/property modifying agents eg ultra-violet absorbers

v   Paints *F* 6

v1   Stopping, fillers, knotting, paint preparation materials including primers

v2   Pigments, dyes, stains

v3   Binders, media eg drying oils

v4   Varnishes, lacquers eg resins
Enamels, glazes

v5   Oil paints, oil-resin paints
Synthetic resin paints
Complete systems including primers

v6   Emulsion paints, where described separately
Synthetic resin-based emulsions
Complete systems including primers

v8   Water paints eg cement paints

v9   Other paints eg metallic paints, paints with aggregates

w   Ancillary materials

w1   Rust removing agents

w3   Fuels

w4   Water

w5   Acids, alkalis

w6   Fertilisers

w7   Cleaning materials F 1
Abrasives

w8   Explosives

w9   Other ancillary materials eg fungicides

x   Vacant

y   Composite materials
Composite materials generally *M* 11
See p. 63 *Construction Indexing Manual*

z   Substances

z1   By state eg fluids

z2   By chemical composition eg organic

z3   By origin eg naturally occurring or manufactured materials

z9   Other substances

Table 4 Activities, Requirements
(main headings only)

Scope: Table 4 identifies objects which assist or affect constuction but are not incorporated in it, and factors such as activities, requirements, properties, and processes.

	Activities, aids
(A)	Administration and management activities, aids, *IC* 2; *M* Introduction, *SF(1)* 2; *SF(2)* 1, 2, 3
(B)	Construction plant, tools *SF(1)* 2; *SF(2)* 2, 11
(C)	Vacant
(D)	Construction operations *SF(1)* 2, 11; *SF(2)* 2, 11

Requirements, properties, building science, construction technology
Factors describing buildings, elements, materials, etc

(E)   Composition, etc *SF(1)* 1, 2; *SF(2)* 1, 2
(F)   Shape, size, etc *SF(1)* 2
(G)   Appearance, etc *M* 1; *F* 6

Factors relating to surroundings, occupancy

(H)   Context, environment

Performance factors

(J)   Mechanics *M* 9; *SF(1)* 3, 4; *SF(2)* 3, 4
(K)   Fire, explosion *M* 1; *SF(2)* 10
(L)   Matter
(M)   Heat, cold *ES* 1
(N)   Light, dark *ES* 1
(O)   Sound, quiet *ES* 1

(Q)   Electricity, magnetism, radiation *ES* 14
(R)   Energy, other physical factor *ES* 7
(T)   Application

Other factors

(U)   Users, resources
(V)   Working factors
(W)   Operation, maintenance factors
(X)   Change, movement, stability factors
(Y)   Economic, commercial factors *M* Introduction; *SF(1)* 2; *SF(2)* 3, 4, 5, 6, 9
(Z)   Peripheral subjects, form of presentation, time, place — may be used for subjects taken from the UDC (*Universal decimal classification*), see BS 1000A 1961

Subdivision: All table 4 codes are subdivided mainly by numbers

Index